ELECTROPLATING OF LEAD FREE SOLDER FOR ELECTRONICS

ELECTRICAL ENGINEERING DEVELOPMENTS

Additional books in this series can be found on Nova's website
under the Series tab.

Additional E-books in this series can be found on Nova's website
under the E-books tab.

ELECTRICAL ENGINEERING DEVELOPMENTS

ELECTROPLATING OF LEAD FREE SOLDER FOR ELECTRONICS

SHANY JOSEPH

AND

GIRISH PHATAK

Nova Science Publishers, Inc.

New York

LIBRARY OF CONGRESS CATALOGING-IN-PUBLICATION DATA

Joseph, Shany.
  Electroplating of lead free solder for electronics / authors, Shany Joseph and Girish Phatak.
   p. cm.
  Includes index.
  ISBN 978-1-61668-753-3 (softcover)
  1. Lead-free electronics manufacturing processes. 2. Electroplating. I. Phatak, Girish. II. Title.
  TK7836.J65 2010
  621.381--dc22
2010013737

Published by Nova Science Publishers, Inc. † New York

Contents

Abstract   **vii**

Chapter 1   Introduction   **1**

Chapter 2   The Changing Face of Soldering
                  in Electronics   **3**

Chapter 3   Options in Solder Materials   **9**

Chapter 4   Electroplating Baths for Promising
                  Lead-Free System   **17**

Chapter 5   Conclusion   **43**

Index   **55**

ABSTRACT

The form and composition of solders used in electronics has undergone many changes throughout the history of electronics. The solder form has changed from wires and ingots to pastes, while the composition has changed from 50Sn/50Pb to eutectic 63Sn/36Pb and a range of intermediate compositions. Two recent developments, *viz.* the health and environmental concerns about lead and the advent of solder bumping and wafer level packaging (WLP) applications in electronic packaging have brought about further changes. Today, electronics needs a reliable process that can deposit solder films with a suitable lead–free composition and form solders bumps up to 10's of micron size upon reflow. Amongst the various options, Sn and its binary and ternary alloys with Cu, Ag,

Bi and In are amongst the contenders for the coveted lead-free solder slot. Co-deposition of these solders in a single electroplating bath is necessary for better reliability of the solder bumping process. However, co-deposition of tin containing lead-free solders by electroplating is a challenging task due to wide difference in the reduction potentials of the individual metals, auto reduction of electropositive elements due to oxidation of Sn^{2+} to Sn^{4+} in the electrolyte and complexities caused by the additives used to control these phenomena. The available literature on co-deposition of binary elements, *viz.* Sn-Cu, Sn-Ag, Sn-Bi and Sn-In, ternary eleme-nts *viz.* Sn-Ag-Cu and Sn-Bi-Cu have been reviewed in detail. While the alkaline baths are discussed, it is the acidic baths which are important and covered in detail due to their compatibility with photolithography. The Acidic baths using sulphuric acid, chloride ions, and methane sulphonic acid (MSA), containing chelating agent, such as, triethanolamine, thiourea, triammmonium citrate *etc* and surfactants and additives such as OPPE *(iso*-octyl phenoxy polyethoxy ethanol), POELE, Polyethylene glycol *etc* are amo-ngst the most reported plating baths for the deposition of above binary and ternary solder films. It is seen that MSA based baths and thiourea as a chelating agent along with appropriate surfactant may form the best option to deposit Sn-Ag-Cu solder film, which is seen as the best lead-free solder option, today.

Chapter 1

INTRODUCTION

Historically, the use of solders dates back to the bronze and iron ages when gold based hard solders were developed by artisans in Mesopotamia, and which very soon spread to Egypt and other places. The use of Sn-Pb alloy traces its origin to the Roman period when 50Sn/50Pb alloy was used for plumbing applications, mainly for joining the lead pipes. Obviously, the same solder alloy was first used in electronics for joining of wires and components [1,2]. The introduction of machines for mass soldering brought about a major change in the alloy composition, because, now, a quickly solidifying solder was required to make hundreds of joints in a few seconds. Thus, around 1950, the eutectic Sn63-37Pb and 60Sn-40Pb compositions came into existence, which continues to be an industry standard even today. Over the years, varying combinations of Sn/Pb alloy ranging from 63Sn/37Pb to 5Sn/95Pb have been used for different applications, even as Sn and Pb continued as the basic constituents of solder. Lead, as a constituent of solder, imparts several advantages to the alloy, such as, improved wetting, enabling intermetallic formation leading to reliable solder joint, prevention of tin whiskers and many others [2]. These advantages have made Sn-Pb the most sought after alloy. However, recently, the environmental concerns due to the toxicity of lead have forced the industry to look for a suitable alternative. Although the search for an appropriate lead-free alloy is on for the past 10-15 years and a range of alloys have been suggested, no alloy has yet been identified as a perfect replacement for the Sn-Pb. This is also due to the fact that electronics itself has been going through series of changes and, consequently, the requirements from the solder alloy with respect to its performance, usage and form, also have been changing. Today solders are required in various forms. Solder wires are needed for manual soldering of wires and components, ingots are required for wave soldering, while the solder paste is required for surface mount Technology (SMT). For the relatively recent solder bumping and attachment process, solder has to be first deposited either by vacuum evaporation using metal/alloy pieces or stencil printed using solder paste or deposited by electroplating the required metal composition. It is seen that amongst these three, electroplating provides a good number of advantages, making it an important contender for obtaining fine solder bumps. The added requirement of lead-free solders, has brought the issues of electroplating of lead-free solders to the fore. At a time when the appropriate lead-free solder is still being finalized, a review of the current status of the electroplating of lead-free solders is not out of place. This is an attempt to review the currently contending lead-free alloys and the efforts to obtain electroplated films of the respective metal compositions.

This paper first reviews how the soldering processes and materials have changed over the decades from hand soldering to the recent flip chip bonding, and from eutectic to high lead solders, followed be the lead-free options. It takes stock of the stringent requirements from the solders with respect to the melting temperature, mechanical properties, the availability and cost, compatibility with the present set-up and summarizes the available lead-free options. The paper takes detailed stock of the available literature including the results obtained by these authors on electroplating of lead-free components and discusses the issues related to their deposition by electroplating. These results are summarized and future trends are discussed.

Chapter 2

THE CHANGING FACE OF SOLDERING IN ELECTRONICS

The process of soldering has undergone a lot of changes over the last century. In the early 20th century, all interconnections were being done with the help of individual solder wires, which were being hand-soldered point-to-point, using soldering irons. After the World War II, there was a huge demand for consumer electronic products and the need for mass production brought forth the new mechanized process in the form of dip soldering and wave soldering, in the late 1950s. The wave soldering technique later found wide acceptance in the industry. Even today, this process is being used due to its ease of operation and lower cost. By 1980s, the circuit complexity and component density multiplied and there was a process that could support high component density and solder them equally quicker. The surface mount technology met these needs. This technology brought in new processes for preparing solder joints, such as vapor phase, infrared, hot gas and other variants of reflow techniques. These changes in processes for soldering brought about changes in the form of solder used. The forms of solder material changed from wires for hand soldering to bulk molten solder for dip and wave soldering and then to a preform or solder paste for surface mount technology. Figure 1 summarizes these changes in soldering technology and in turn, the changes in the forms of solder over the years. Significantly, throughout the above changes in the process and form of solder being used in electronics, the eutectic Sn-Pb solder remained the favorite choice [2]. Its low eutectic melting point with a unique combination of favorable properties along with low cost made the alloy almost indispensable.

In 1964, IBM introduced a new process called Controlled Collapsible Chip Connection (C4), using solder bumps. These solder balls or bumps are developed on the conductor pads by depositing solder of appropriate volume and the bump are obtained after reflow of the solder. Such bumped chip can be attached to the substrate by simple reflow, as in the case of surface mount devices. This method opened up an all-new application for the soldering. Now, solders were replacing the pins of the package as well as the wire bonding of chip inside the packages. This new method of interconnections was an area array process, which could accommodate a larger number of interconnects at finer pitches. This reduced the size of the packages dramatically since area array interconnections allowed very high interconnection density. Figure 2 schematically shows such solder bump interconnections.

The preparation of solder bumps on the chip is facilitated by a series of metal layers below the bump called Under Bump Metallurgy (UBM). A similar structure is many-a-times developed on the substrate where these bumps are eventually attached. These layers on the substrates are called Top Surface Metallurgy (TSM). The main functions of these layers on either side of the solder is to provide adhesion, control unintended and undesirable diffusion of pad metal into the molten solder and provide a wetting layer for the solder [3]. Various UBM have been reported in the literature. The commonly used UBM for Sn-Pb systems include: Cu/Cr/Au, TiW/Cu/Cu, Cr/CrCu/Au, Ti/Cu (PVD)/Cu (electroplated), Electroless Nickel (EN)/Au, Cr/Cu/Au/Sn, Cu/EN [4] Cu/Ta/Cu and Ti/Cu, NiV/Cu, Ni/Au [5,6] Cr/Cr-Cu/Cu [7] *etc.* Some of these layers are also being studied for lead-free systems.

These are usually sputtered, electroplated or deposited by electroless technique. Amongst lead-free systems, electroless Ni/Au layers as UBM have been studied for stencil printed Sn4Ag0.5Cu and Sn 3.5Ag alloys [8]. A low cost UBM comprising of Electroless Ni / immersion gold (EN/IG) has been reported for C4-New process (C4NP)[9]. It is reported that for high Sn containing alloys, Cr/Cr-Cu/Cu have shown poor results. This is due to the high Sn content of eutectic solders consuming the Cu from the Cr-Cu matrix, which results in a weaker interface [10, 11].

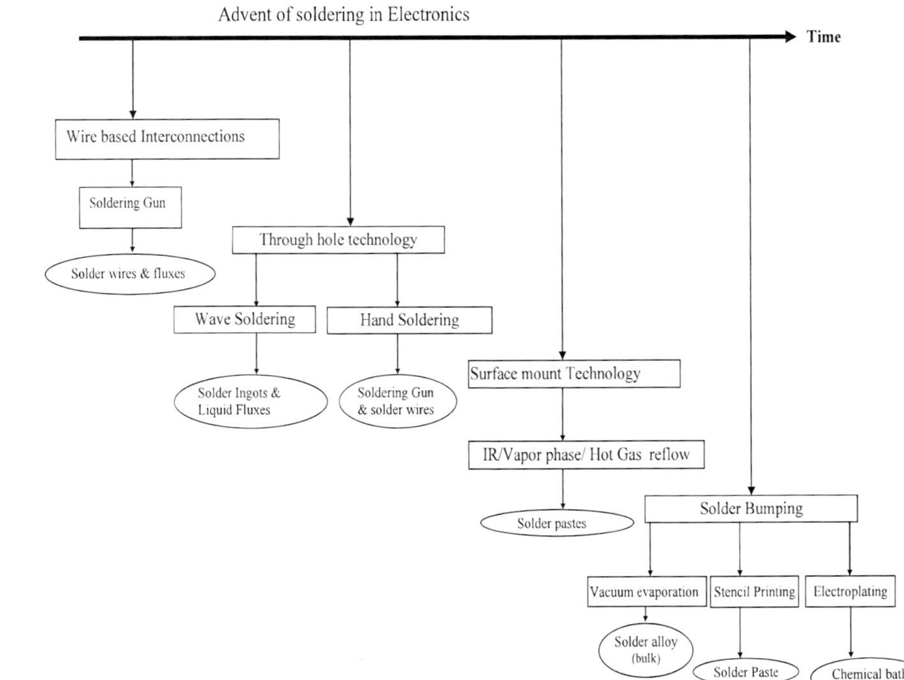

Figure 1. Evolution of various soldering technologies and solder forms in the electronic industry

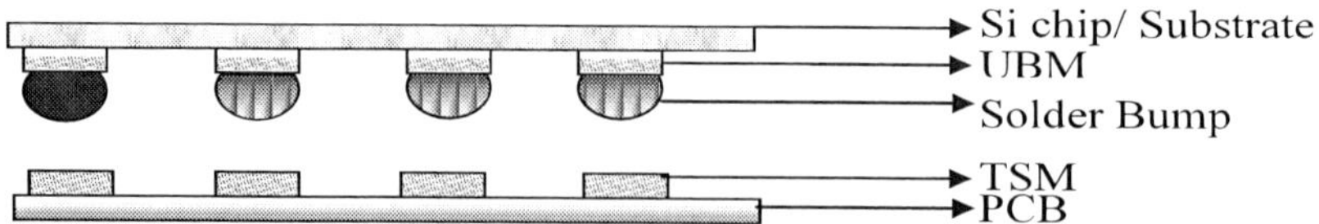

Figure 2. Schematic showing. solder bump attachment

Just as UBM, TSM can also be formed by either thick or thin film technology [12]. A combination of materials, such as, Cu, Pd, Pt, or Ni is used to form TSM [7]. The selection of materials for UBM and TSM largely depends upon the choice of the solder bump materials. An appropriate selection of UBM and TSM is critical to the reliability of solder bump. Overall, the following functions have to be carried out by these layers:

- It should act as diffusion barrier layer to prevent dissolution of pad material into the molten solder. Layers of Ti, TiW, and Cr in the UBM function as diffusion barrier layer.
- It should act as adhesion enhancing layer, for which layers of copper, nickel or nickel vanadate can be used.
- There may be an optional oxidation barrier layer, such as, gold or palladium for preventing oxidation of final UBM layer.

All the solder bumps on a chip or package are developed simultaneously, and this process must ensure that the bumps have uniform physical and chemical properties. Any mismatch in the dimensions or the chemical composition of the bumps would raise concern over the reliability of the final joint. If case of such non-uniformity, some of the bumps may not melt completely during reflow and cause a weak solder bond that may give away during handling. However, with the advancements in the deposition processes, it is now possible to control the properties of these high-density interconnections.

Several processes have been demonstrated for the deposition of solder bumps on a wafer or substrate. These include evaporation, stencil printing, electroplating and dispensing. Primarily, the bump dimensions, pitch, composition, and the cost influence the choice of one technology over another. In order to realize the evaporated solder bumps, IBM used high lead containing Pb/Sn alloys, which were deposited by evaporation through metal mask. This technique has limitation with respect to the minimum pitch size. It is difficult to use this technique for bump pitches lower than 225μm. Solder paste deposition using stencil printing followed by its reflow to form bumps is another user friendly process that can be used for obtaining high density solder bumps. With fine solder paste (Type VI: 20–25μm or +500 – 635 mesh) the finest pitch possible is about 175–200μm [13]. The rheology of the paste, the print speed, the aspect ratio of the pattern and snap-off (the separation between the stencil and the substrate during printing) are some of the factors that affect the printed bump geometry and uniformity. In general, bump pitches smaller than 175μm are difficult to obtain using this technique. Both the above-mentioned processes have the basic infrastructure available and have shown good reproducibility within their own size and pitch limitations. While evaporation process has a long manufacturing history and good reliability, the stencil printing process is cheaper. Composition control for low melting solder constituents is another issue that needs to be handled when depositing metal alloys by evaporation technique.

In contrast, electroplating possesses all the ingredients to become a popular alternative to the evaporation process because of its lower equipment costs, minimal facility and floor space requirements and low cost of processing. The electroplating process is used in combination with additive or subtractive photolithographic processes to obtain solder deposition on pre-defined area, although the additive process is most common. The photolithographic processes often determine the geometry and pitch limitations of plated bumps. This being a proven process, its extendibility in terms of progressively smaller bump dimensions is excellent. Additionally, electroplated bumps offer many advantages, including, lack of voiding, microstructure and thickness control, possibility of high temperature solder deposition, manufacturability, and good composition control. The electroplating process for solder bump preparation is also cheap, comparatively more flexible and provides finer bumps and pitches [7]. Use of suitable salts and additives makes it possible to deposit solders with desired composition using electroplating. On the flip side, the electroplating processes could be cumbersome, considering that these processes may need surface pre-treatment before plating [14]. Nevertheless, this disadvantage is quite minor as compared to the cost and small-dimension advantage available from electroplating.

In the electroplating of multi-element system, the solder alloy is expected to form only after the reflow of the deposited solder. Therefore, the electroplating processes should ensure presence of the solder metal components in the desired composition on a wetting surface. This gives rise to the two ways of electrodepositing a multi-element system; *viz.* co-deposition and sequential deposition. In co-deposition, all the elements to be plated are present together in the plating solution, and by means of suitable chelating agents and additives, and by controlling the deposition conditions, these elements are deposited simultaneously in the desired composition. On the other hand, in sequential process the desired elements are plated individually using single component baths. A reflow process is followed after plating in each case, which forms the desired solder alloy bump. Amongst the two, co-deposition is seen as the best option with respect to the solder composition control even though it faces difficulties, such as, difference in electrode potential of elements, and issues related to bath stability. In the case of sequential depositions, a difficulty in plating could be the replacement reaction during electro-plating, which may complicate the deposition process. Precise thickness control of individual layers is most crucial for the sequential plating process; else the bump composition may vary. This is especially true when a particular composition requires small quantity of a particular metal and when the bump dimensions are small. Overall, the co-deposition process offers many advantages, only if an appropriate bath for co-deposition is identified.

Chapter 3

OPTIONS IN SOLDER MATERIALS

As one of the primary components of the solder, lead imparts many advantages to tin-lead solders. It reduces the surface tension of tin, thereby facilitating wetting. As an impurity in tin, it prevents the transformation of white (or beta tin) to gray (or alpha tin) during cooling. Such a transformation would otherwise cause loss of structural integrity of the solder. Lead prevents the formation of tin whiskers. Lead also serves as a solvent metal, enabling the other joint constituents, such as tin and copper, to form intermetallic bonds rapidly. These factors, in addition to its easy availability and low cost, make Pb an ideal alloying element with tin for soldering applications [2, 3].

In the recent years, however, many health hazards caused by lead have come to the fore. Blood lead level (BLL) exceeding $10\mu g/dL$ is hazardous for health. In adults, lead poisoning can cause various symptoms, including fatigue, loss of appetite, stomach disorders, hypertension, anemia, dizziness, and weakness in the extremities [3]. The hazards are much more serious when children are exposed to lead. Lead poisoning in children can cause learning disabilities, attention deficit disorders, and lower IQ. Elevated levels of lead sustained over a period of time can damage the central nervous system of children and adversely impact their development. The toxic effects of lead, and hence, their applications in electronics have been in serious discussion since the 1990s. There was a call for a ban on its use, and the phasing out was proposed to be carried out in stages. World-wide, about 5 million tons [3] of solder is being used annually in electronics. This forms just about 1% of its total use, but its hazardous impact may be much higher due to the way the electronic scrap is handled. Discarded electronic products remain in the landfills causing leaching of lead into the ground water. The concern about use of lead in electronics industry also stems from its occupational exposure. In the through-hole assembly process, the dross or oxides formed in the solder pots can be the greatest routes for lead exposure to the operator since it has to be removed periodically by him. In the SMT process, solder paste waste, stencil wipes, disposable gloves, stencil cleaning by-products, and board washing solutions are all potential sources of lead contamination to the environment. The industry is still undecided over the use of lead-free systems for high reliability applications in military and space systems, but the manufacturers in the consumer electronics sector are now looking for an appropriate lead-free alternative.

With the eutectic 63Sn-37Pb as the benchmark, various lead-free options are being explored for an appropriate replacement of the lead based solder. It is now clear, however, that an exact 'drop-in' replacement is not possible. The Sn-Pb solder has a wonderful cross-section of properties in terms of the melting point, wetting, whisker prevention, mechanical strength, reliability and cost. Even after almost two decades of search for the lead-free solder materials, researchers have not been able to find an alloy that has similar, if not better, set of properties. The industry demands a low cost alloy that can fit in the present industrial set-up without any major change in the processing conditions. Table 1 presents the important properties of the alloy that need to be considered while selecting an appropriate solder alloy, along with the corresponding Sn-Pb solder properties

Unfortunately, there is no single alloy that can be simply called a one-for-one replacement for tin/lead eutectic. Along with Sn, elements like Cu, Ag, Bi, Zn, In, Sb & others in binary or multiple combinations could be used [3,15]. Various compilation about the properties of the Sn containing lead-free options have identified a small group of alloys which show properties close to the ideal ones, and have less compromises. The following sub-section presents a summary of the properties of these individual alloy systems.

Table 1. Important properties to be considered while selecting the lead-free alloy

S. No.	Properties	Needs from Lead-free alloy	The Sn-Pb Benchmark
1	Melting point	Determines the soldering temperature i.e. the temperature to which the PCB will be raised during soldering. Reflow temperature should be less than 250°C for polymeric PCBs. Melting pint of the solder must be preferably below 225°C	Eutectic melting point of 183°C and reflow temperature of 210-215°C
2	Mechanical strength and reliability	Good physical and mechanical properties like tensile strength, shear strength, reliability during thermal cycling and vibration and mechanical shocks, good wetting, corrosion resistance, matching thermal coefficient of expansion etc.	Tensile strength:40Nmm^{-2}, Shear strength : 23 Nmm^{-2}, Thermal expansion coeff. : 25ppm/$^{\circ}$K, No. of cycles to failure: 3650
3	Environmental concerns	Should be environment friendly and should not cause any sort of potential health hazard	Pb is toxic and causes potential health hazard
4	Electrical Resistivity	As low as possible. Very important for electrical signal transmission	14.6 $\mu\Omega$.cm
5	Good thermal conductivity	As High as possible. Required for efficient thermal dissipation	50 W/mK@25°C
6	Easy reparability	Repair/rework requires simultaneous melting of all joints and restoration of these joints after rework. Further, the alloy properties must be retained after re-work	Sn63/37Pb being a eutectic alloy is easy to rework
7	Cost	The alloy should be cost effective.	The a cost of eutectic Sn-Pb is about \$2.5/lb
8	Availability	The alloy should be easily available in sufficient quantities at reasonable cost	Easily available
9	Compatibility with the present set-up	The selected alloy should be compatible with existing and future materials and processes to allow smooth transition	The industry is set-up for Sn-Pb

3.1. Tin

Tin has an ability to wet and spread on a wide range of substrates using mild fluxes [14, 16]. It has a melting point of 232°C, which means that the PCB would require reflow at temperatures beyond 260°C. Apart from this, tin has a tendency to transform to grey tin (from 'β' phase with body centered tetragonal structure to 'α' phase with diamond cubic structure) at temperatures below 13°C, which may induce cracks in the tin structure due to the increase in volume [16]. Tin also has a tendency to form whiskers.[1] Although whiskers do not affect the soldering capability or deteriorate the tin coating, but may cause electrical shorts in PC board assemblies [16], which may prove to be critical at small dimensions. Adding small amounts of metals such as Pb, Co or Sb can reduce these difficulties [14].

3.2. Tin-Copper

With a eutectic composition of Sn-0.7Cu, this is one of the commonly used solder alloys for plumbing applications. This is one of the cheapest (~1.5 times of Sn-Pb cost) amongst the lead-free alloys available and is compatible with lead bearing finishes. In wave soldering process this alloy is preferred mainly due to the need of frequent replacement of the entire soldering material, and the low cost of Sn-Cu makes it affordable [3]. The melting point of the alloy is 227°C which is on the higher side for electronics applications. Presence of copper in the alloy itself restricts the dissolution of copper from the pads. However, this alloy also has a problem of whisker formation due to high tin content. Further, this alloy is probably the poorest in mechanical properties of all available lead-free solders [18].

---

[1] Whiskers are needle like crystals projecting from the surface of a tin layer with diameters of the order of a few micrometers and length of the order of few millimetres. This problem is usually found in electrodeposited tin layers containing certain amount of residual stress. To relieve the stress in the electrodeposited layer the atoms move into a hair thin crystal. [14]. These are usually straight or kinked with straight segments between kinks. Some recent work suggests that tin whiskers

3.3. TIN-SILVER

The tin-silver eutectic alloy (Sn-3.5Ag) has a liquidus temperature of 221°C. This is the most reliable alloy with the longest history of use amongst lead-free alloys [18]. It shows excellent mechanical properties and better solderability than tin-copper. Silver being an expensive metal, it increases the cost of the alloy. Again, the high tin content may cause whisker growth. Further, Sn-Ag alloys exhibit poor wetting characteristics [18]. Addition of Bi (<5%), and In helps in improving this property of the alloy. In fact, Sn-Ag-Bi shows the best solderability amongst a range of lead-free alloys [18].

3.4. TIN-ZINC

Zinc in solder alloy helps in reducing the melting point. Eutectic Sn-9Zn has a melting point of 198°C. However, the use of zinc has been limited due to its low chemical stability, tendency to oxidize in air, and dross formation. Zinc is prone to corrosion and may have some wetting issues. Due to the oxidation problem of zinc, specialized fluxes and adoption of special soldering process conditions are required to achieve acceptable solder joints. Addition of Bi to Sn-Zn helps in eliminating the corrosion problems. Zinc alloys also have embrittlement problems [15, 19, 20].

3.5. TIN-BISMUTH

The tin–bismuth alloy is said to have characteristic closer to Sn-Pb. This solder is less stressful to PCB and the components and also has a proven reliability record in high-end computer applications [15]. The eutectic binary alloy-42Sn-58Bi has a melting point of 138°C. However, Sn-Bi is susceptible to brittle fracture under shock loading and needs alloy modification to reduce its strain rate sensitivity. The main drawback of bismuth alloy is the formation of a low melting (96°C) Sn-

---

are capable of carrying current between 20 and 30mA under a voltage of 15V before burning out [17].

Pb-Bi ternary phase in presence of lead [15, 18]. During the thermal cycling tests between 0-100°C, this phase falls out, thus destroying the solder joint. Since the solubility of Bi in Sn is just around 2% at room temperature, at Bi contents of ~ 2.5% or higher Bi may precipitate out by prolonged low temperature exposure [19]. These precipitates could form grain boundary film, which could be disastrous on sudden heating to about 138°C. This makes it necessary to restrict Bi content to less that 5% and preferably below 2%, even though Bismuth increases the strength of the alloy and reduces its ductility [19]. Addition of bismuth reduces the melting temperature of the alloys and improves the mechanical properties, such as, tensile strength, yield strength, Young's modulus *etc.* When 1%Bi is added to eutectic Sn-Ag, it exhibits fatigue life significantly higher than Sn-37Pb and the eutectic Sn-Ag. Bismuth based alloys containing 2-5% Bi could be a possible alternative for Sn-Pb. More studies in this direction are required before reaching a final conclusion because Bismuth is a by-product of lead mining, which could bring forth availability issues.

3.6. TIN-SILVER-COPPER

The Sn-3.8Ag-0.7Cu eutectic has a melting point of 217°C. Incorporation of Cu in Sn-Ag alloy helps in reducing the Cu dissolution from pads, lowers the melting point, improves wetting and reduce screep and thermal fatigue [18]. This alloy shows better reliability and solderability as compared to Sn-Ag and Sn-Cu and seems to be one of the most promising amongst the lead-free alloys [3, 18]. A minor addition of metals, such as, indium or antimony, improves the wetting capability of the alloy. Bismuth & cobalt, added in small quantities, helps in improving the mechanical properties of the alloy [15, 21].

3.7. TIN-SILVER-BISMUTH-COPPER

The alloynSn-3.1Ag-3.1Bi-0.5Cu reportedly exhibits best balance of desirable properties like high fatigue resistance, high strength with sufficient plasticity [21]. This alloy has a melting range of 209-212°C, which is about 10°C lower than the ternary Sn-Ag-Cu eutectic [21]. The properties, such as, tensile strength, Young's modulus, number of cycles to failure, were found to be much higher than the Sn-Pb or other common lead-free binary alloys. The presence of Bi is the only cause of concern in this alloy. Formation of low melting phase in the presence of Pb would be an issue during thermal cycling. Additionally, since four elements are involved, their composition control should be looked into carefully.

3.8. OTHERS

Some of the other elements that can be used to make solder alloys are: indium, antimony, gallium, mercury, cadmium *etc.* Indium has a low melting point (157°C) and the binary eutectic Sn-48In has still lower melting point of 118°C. Although it has good oxidation resistance, its susceptibility to moisture induced corrosion, softness, low strength and high cost are the major reasons for the limited applications of indium and indium-based alloys. Antimony in alloys improves the thermal fatigue properties. Further, the addition of antimony as a dopant in Sn-Ag-Cu alloys reduces the melting temperature and refines the grain structure lightly [20]. There is a good deal of concern and confusion regarding the toxicity of antimony. Even gallium, cadmium and mercury have compelling disadvantages to cause any serious consideration amongst the lead-free options. For example, cadmium is undesirable as it is toxic, while gallium has a limited supply, and is very costly. It also causes embrittlement problems and has a low melting point [19, 20].

The above discussion provides the summary of different possible lead-free alloys in consideration. While the conclusions are slow, it is seen that the discussion and work in the area of lead-free solders and their deposition by electroplating is increasingly narrowed down to Sn alone or binary, ternary or quaternary compositions of Ag, Cu, Bi and Zn with Sn. Indium is being ruled out for its high cost and low strength apart from moisture induced corrosion problems. While the advantage of the Sn-Bi and Sn-Zn binary eutectic is their lower melting points, they are more reliable in their ternary combinations with Ag and Bi. However, as Ag content increases, their liquidus temperature and the pasty range also increases [22}. As the composition of Bi in the alloy increase, its ductility reduces and the alloy tends to become brittle. To sum-up, Table 2 presents important properties of the common eutectic alloys of the mentioned elements with Sn. Some of them could be directly in contention for the lead-free alloy slot, while the off-eutectic compositions of some of these may be of importance.

Table 2. Physical and mechanical properties of some selected solder alloys [2, 3, 15, 21, 22]

S. No.	Alloy Property	Sn-37Pb	Sn	Sn-3.5Ag	Sn-0.7Cu	Sn-3.5Ag-0.8Cu	Sn-3.1Ag-3.1Bi-0.5Cu	Sn-58Bi	Sn-9Zn
1	Melting Point	183°C	232°C	221°C	227°C	217°C	209-212°C	138°C	193°C
2	Electrical resistivity ($\mu\Omega$.cm)	14.6	11.2	12.3	$10-15$	13	-	$30-35$	$10-15$
3	Thermal conductivity (W/mK)@25°C	50	66.6	78	65	60	-	19	-
4	Tensile strength N/mm2	40	-	48		48	79	-	-
5	Joint shear strength, N/mm2 at 0.1mm/min	23	-	27	23	27	-	-	-
6	Young's Modulus (Gpa)	27	-	44	26	-	45	-	-
7	Plastic strain at fracture (%)	24	-	38	44	-	18	-	-
8	No of Cycles to fracture	3650	-	4186	1125	-	5923	-	-
9	Cost	2.37	-	6.32	3.48	6.55	-	3.44	3.23
10	Toxicity	Yes	No	No	No	No	No	No	No
11	Whisker formation	Nil	High	low	medium	low	Nil	Nil	Nil

Chapter 4

ELECTROPLATING BATHS FOR PROMISING LEAD-FREE SYSTEM

An ideal co-deposition bath is one, which is able to deposit film with the desired composition range uniformly across the deposited area, and has equally uniform and compact microstructure. A plating bath is reliable when it produces consistent results at identical condition. It is better when all the elements deposit at a single or close range of potential. In fact, a desirable electroplating system for co-deposition is one where the difference in reduction potential between the co-depositing metal ions is less than 0.2V [2]. Unfortunately, this condition is not met in any of the co-deposition baths containing Sn and other selected elements. This is because, amongst Sn, Cu, Bi and Ag, Sn is a common element which has negative reduction potential while the rest, except Zn, have positive reduction potentials which are wide apart from Sn. The reduction potential of Zn is more negative than Sn in the electrochemical series. Table 3 presents the difference in reduction potentials for these elements with Sn. For a practical co-deposition bath, therefore, one or more chelating agents have to be identified which can help in co-deposition of all the elements in the desired composition at a close deposition potential. These additives however, may complicate the bath. It is absolutely necessary that all the bath constituents are chemically compatible so as to ensure a stable bath. This is especially true in a multi-element system, where there is possibility of the metal salts or their chelates and other additives reacting to form side products which may be undesirable [2]. These requirements of a plating bath are common for all the co-deposition systems, including those discussed here. The following subsections discuss the efforts to discover such practical plating baths for the identified lead-free systems.

Table 3. Difference in electrode potential of the common lead-free alloy systems

S. No.	Alloy System	Difference in reduction potentials (V)
1	Sn-Ag	+0.926
2	Sn-Cu	+0.476
3	Sn-Bi	+0.446
4	Sn-Zn	-0.624

4.1. TIN

In solutions, Sn exists in two forms *i.e.* divalent Sn^{2+} and tetravalent Sn^{4+}. Both these forms are stable and convertible into each other. In acidic solution, Sn is present as Sn^{2+} ions and in the alkaline range, mainly above the pH of 13, Sn^{4+} is more prevalent. At intermediate pH range there could be conversion from one form to other and equilibrium is soon attained. Most of the baths in commercial use are acidic and the Sn^{2+} ions readily gets oxidized to Sn^{4+} in presence of atmospheric oxygen. During electroplating of pure Sn or the high tin containing alloys, it should be ensured that oxidation of Sn^{2+} to Sn^{4+} is minimal. Presence of reducing agents in the plating baths would help in controlling the oxidation of Sn^{2+}. Another important issue in tin deposition is to optimize the deposition parameters to obtain smooth, uniform and stress free-film. Any residual stress in the deposited film leads to the formation of whiskers. Adding small amounts of metals such as Pb, Co, Bi or Sb can reduce these difficulties [2,3]

The most common plating bath systems for Sn are those based on fluoboric acid, sulfuric acid, methane sulfonic acid, hydrofluoric acid and phenol sulfonic acid. The fluoborate acid baths are one of the oldest baths for Sn plating which can take up high current density ($300mAcm^{-2}$) and have good throwing power, but environmental concerns due to fluoride and boric ions and their effluent treatment issues, have restricted their use. The most commonly used organic additives in this system include peptone, gelatin, β-naphthol. Catechol and hydroquinone are used as antioxidants [2]. The sulfuric acid based baths have been long in use, owing to their low cost and high throwing power. Gelatin, β-naphthol, resourcinol, cresol-sulfonic acid or phenol are amongst the most commonly used additives in this bath system. Anode passivation at high current density and corrosion of the plating equipment due to the plating solution are the major drawbacks of this bath [2]. The methane sulfonic acid (MSA) based plating system is the current favourite among the electroplaters. The advantages of this system include its excellent metal salt solubility and low corrosivity. MSA based baths are environmentally friendly due to their low toxicity, ease of effluent treatment, and biodegradability [2, 23] Glycol additives in methane sulfonic acid baths help in reducing the hydrogen evolution at low over potentials [24] and refine the grain structure over a wide current density range. This behavior is more effective in presence of phenolphthalein. Perfluorinated cationic surfactant is also reported to function as a primary additive and control hydrogen evolution [23]. Hydroquinone is a very common antioxidant and is found to be effective in methane sulfonate bath.[2, 25] The additives in the bath include those necessary to improve the surface morphology, antioxidants to control the oxidation of Sn^{2+} to Sn^{4+}, surfactants to promote electrode reactions, bath stabilizers and some others. The alkaline bath might require a different combination of bath additives. In the presence of bath additives, the deposition potential of Sn^{2+} (-0.136V) may shift to either more positive or negative values. Selecting appropriate bath constituents is, therefore, very critical to obtain the required film properties.

Amongst the alkaline baths, sodium and potassium stannate baths are commonly used. The potassium stannate bath is in more demand due to its better solubility and its ability to take higher current density. The alkaline baths are reported to have better throwing power [2]. For best output, it is critical to control the alkali concentration. Owing to the problem of whisker formation, pure tin solder depositions are usually avoided.

4.2. Tin-Copper

The eutectic composition of Sn-0.7Cu has a melting point of around 227^0C. The electrodeposited Cu-Sn alloys have been reported to show low surface tension, malleability, ductility, solderability or resistance to corrosion, depending upon their composition [26]. The electro deposition of Sn-Cu is complex owing to the wide difference in the reduction potentials and the oxidative tendency of Sn^{2+} to form Sn^{4+} that leads to the reduction of Cu ions in the electrolyte itself. Obviously, a good combination of antioxidants and chelating agents is a must for the co-deposition of Sn and Cu. Additionally, additives may be necessary for improving the film microstructure and the compositional uniformity of the deposited films.

Since long, copper plating is being done in the industry using a cyanide bath, sulfate or fluoborate based acid baths or pyrophosphate based alkaline baths. In Sn-Cu systems there are reservations on the use of cyanide based baths owing to its high toxicity and hence, difficulty in handling and in effluent treatment to satisfy the environmental norms. Non-cyanide baths, both acidic and alkaline are, therefore, are being explored for commercial use. Here, it is important to note that the acidic baths have an edge over alkaline baths especially where photolithography and film patterning is required. Most of the commercially available photoresists are resistant to acidic baths and dissolve in the alkaline electrolytes. There are quite a few reports in the literature on the various electrolyte systems and additives and chelating agents used for the electro deposition of Sn-Cu alloy. Table 4 presents a summary of some of the reported baths for electro deposition of Sn-Cu alloys.

The most commonly used acidic plating baths are those based on sulfuric acid or methane sulfonic acid. Chloride based electrolytes with citrate as complex forming ion is also reported. Tartrate and citrate complexes with Sn as well as copper are widely reported [36-38]. However, it is also reported that with solution aging, copper ions reduce easily from the Cu-citrate complexes, which causes slight increase in copper content in the deposits [39]. Methane sulfonic acid is itself a reducing agent and controls the oxidation of stannous ion thereby ensuring enhanced bath stability.

Table 4. Summary of reported plating baths for the co-deposition of binary Sn- Cu films

S no	Bath type	Bath constituents	Plating condition	Bath/Film properties	Remarks	References
1	Acidic	Sulfuric acid based, $SnCl_2$, $CuSO_4$, Sodium Potassium Tartarate	Temperature:25°C Current Density: 11mAcm^{-2}	Tartarate inhibits vertical growth and promotes crystallite growth rate. Refined film morphology. Compact and uniform film microstructure.	Tartarate helps in improving the bath stability by inhibiting the oxidation of Sn^{2+} to Sn^{4+}. Bath retained its deposition efficiency even after 4 weeks	[26]
2	Acidic	Chloride based bath containing $SnCl_2$, $CuCl_2$`, Triammonium citrate (TAC)	Temperature: 25°C Time: 30 min. Current density: 2.5-20mAcm^{-2} pH: 4-5	Near-eutectic Sn-Cu films at 10mA/cm^2. plating rate: 20µm/hr	Improved solution life, TAC functioned as bath stabilizer. Solution showed good stability for more than 36 days.	[27,28]
3	Acidic	Sulfuric Acid Based electrolyte, $SnSO_4$, $CuSO_4$, Thiourea , Polyoxyethylene Lauryl Ether (POELE)		Thiourea is used as chelating agent. POELE as an additive gets adsorbed on the deposited Sn film and inhibits further deposition thus producing a surface smoothing effect.	---	[29]
4	Acidic	Sulfuric Acid Based electrolyte, $SnSO_4$, Cu_2O, thiourea, Pyrocathecol	Current density: 10-100mAcm^{-2}Temperature: 30-50°C electrolyte stirring	Thiourea functioned as chelating agents. Sn-4Cu films obtained at 40°C using 100mAcm^{-2}.	---	[30]
5	Acidic	Commercial bath from MacDiarmid Co containing MSA, Sn & Cu salts	Temperature:20°C Current density: 10-80mAcm^{-2}	Film deposition rate- 0.25-2.7µm/min. Film composition- Sn- (0.9-1.4%)Cu	This bath was used to obtain 140µm Sn-Cu solder bumps with a bond strength of 113gf	[31]
6	Acidic	MSA Based commercial bath, methane sulfonate salt of Sn & Cu, commercial additive FCM1 & FCM 2	Current density: 200mAcm^{-2} Temperature: 40°C	Additives help in controlling the internal stresses in the deposited film and prevent whisker formation	Sn-2Cu solder when tested on packages for reliability was found to be free of whisker growth, solder cracking. Showed reliable solderability and robust solder joint.	[32]

Table 4. (continued)

S no	Bath type	Bath constituents	Plating condition	Bath/Film properties	Remarks	References
7	Acidic	Sulfuric Acid Based electrolyte, $SnSO_4$, $CuSO_4$, Commercial surface active agent Laprol 2402C	Current Density: 2-25mAcm^{-2}	Additive gets adsorbed on the deposited film and inhibits/ controls further deposition.		[33]
8	Alkaline	Non-Cyanide bath. $CuSO_4$, Sn Stannate containing NaOH and Sorbitol	Electrolyte stirring Temperature-25°C PH: 12.7	With pulse plating Cu content varied from 1.5-3.5wt% and DC plating Cu content varied as-1.5 to 2.5wt%	Sorbitol functioned as chelating agent, brightener and leveler.	[34]
9	Alkaline	$K_4P_2O_7$, $Sn_2P_2O_7$, $Cu_2P_2O_7$, a Brightener, a Cationic surfactant Surface tension adjusting agent and MSA	pH : 8.01 Temperature: 25-30°C Current density: 5-30mAcm^{-2}	Glossy, silver-white beautiful film, Film composition: 99-30Sn/1-70Cuwt% Film had good solder wettability	MSA functioned as bath stabilizer	[35]

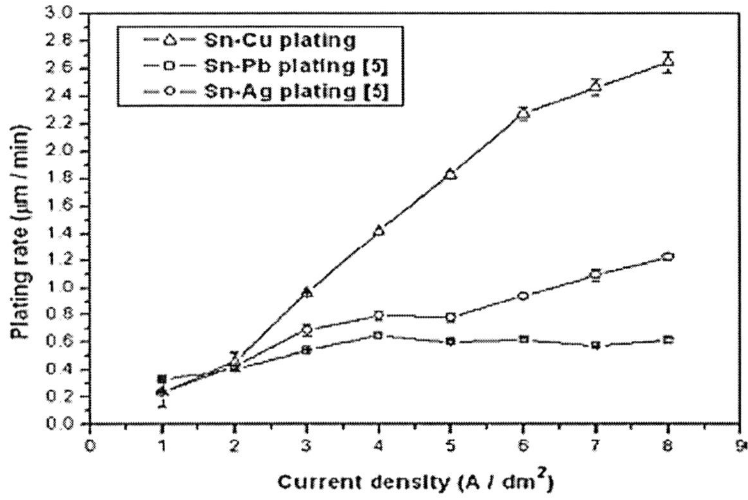

Figure 3. Dependence of Deposition rate of solders on current density[31, Reprinted from *IEEE Trans. Elec Pckg Mnfg* , 29, No.1 , Seok Won Jung, Jae Pil Jung & Y (Norman) Zhou, ''Characteristics of Sn-Cu solder bump formed by electroplating for Flip Chip', 10-16, Copyright (2006), with permission from IEEE]

Figure 4 Sn–Cu solder bumps formed on a Si-wafer by electroplating [31, Reprinted from *IEEE Trans. Elec Pckg Mnfg* , 29, No.1 , Seok Won Jung, Jae Pil Jung & Y (Norman) Zhou, ''Characteristics of Sn-Cu solder bump formed by electroplating for Flip Chip', 10-16, Copyright (2006), with permission from IEEE]

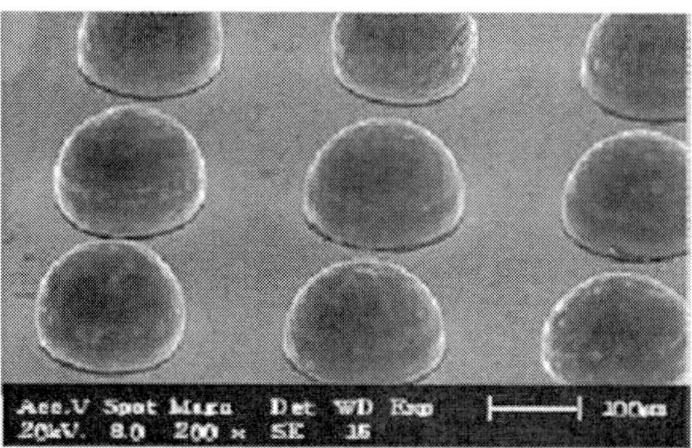

Figure 5. SEM Micrograph of Sn–Cu solder bumps after reflow reflow [31, Reprinted from *IEEE Trans. Elec Pckg Mnfg* , 29, No.1 , Seok Won Jung, Jae Pil Jung & Y (Norman) Zhou, ''Characteristics of Sn-Cu solder bump formed by electroplating for Flip Chip', 10-16, Copyright (2006), with permission from IEEE]]

Sometimes this removes the need of any additional antioxidant. In general, the MSA based bath can take high current density and has shown high film growth rate. The bath would be slightly costly as compared to the sulfuric acid bath. Thiourea is very effective as a chelating agent and its complex with copper has been reportedly utilized for many applications, such as, electroplating, phase separation, biological applications *etc* [40-47]. The complex formation can be utilized to control the composition of copper during the binary deposition. The reports confirm its utility in both sulfuric acid and MSA based systems. POELE, TritonX100$^{®}$, tartarate and other non-ionic surfactants, various commercial additives, triethanol amine, sodium gluconate, 1-4 hydroxybenzene, polypropelene glycol *etc*. [28] have been used to improve the properties of the deposited film as well as to enhance the bath stability. Seok W. Jung and his group [31] have used a MSA based commercial Sn-Cu bath from MacDiarmid Co. for obtaining solder bumps. The electroplating was carried out at 20°C using DC current and the electrolyte was agitated at 240r/min. The plating rates at different current density are presented in Figure 3. The plating rates are compared with the Sn-Pb and Sn-Ag deposition carried out by Hwang and his group [48]. It is seen from this figure that the MSA based baths offer high film deposition rates, up to 2.7μm/min. In comparison, the other baths have shown deposition rates much below 1µm/min. This bath was used to deposit Sn-Cu solder films on patterned substrates with110µm diameter pads with 250µm pitch. At 50mAcm^{-2} and a plating time of 2 hours, the mushroom shape depositions were obtained. Figure 4 and 5 reproduce the micrographs of as-deposited Sn-Cu films as well as solder bumps obtained after reflow at 260°C respectively.

Amongst the alkaline baths, pyrophosphate bath is the most commonly studied non-cyanide bath. It uses sorbitol as a chelating agent, which also helps in improving the film morphology. In stannate baths, sorbitol functioned as an effective chelating agent, brightener and leveler. It gave good stability to the bath at a pH of ~12.7 and improved the microstructure. MSA was used as bath stabilizer in a pyrophosphate bath which was maintained at a pH of 8 to obtain a silver-white Sn-Cu film with good wetting properties.

4.3. TIN–SILVER

There are very few reports on electroplating of Sn-Ag, probably because of the difficulty in their co-deposition due to the wide difference in their deposition potentials. Most of the early work on Sn-Ag deposition was targeted at replacing pure silver coatings with Sn-Ag coating containing about 10-15% Sn to reduce the tarnish of the components with pure silver. The most common plating bath for this binary system is cyanide based. Cyanide ion is a very effective chelating agent for silver and forms part of the most reliable plating system. The toxic effects of cyanide have driven the exploration of other chemical combinations that can effectively plate the desired alloy combination. Table 5 presents the summary of various reported baths for Sn-Ag deposition.

For the non-cyanide and acidic baths, thiourea appears to be the most studied chelate for silver depositions and can be used for sulfuric acid and MSA based systems. Although silver complexes with cyanide and thiosulfate are quite stable and widely used, this complex cannot be used in MSA based electrolytes because of reactions between MSA and silver complexes [2].

Table 5. Summary of reported plating baths for the co-deposition of binary Sn-Ag films

S no	Bath type	Bath constituents	Process condition	Bath/Film properties	Remarks	Reference
1	Acidic	H_2SO_4 based, $SnSO_4$, $AgNO_3$,thiourea, commercial surface finishing agent	Temperature: 25°C Current Density: 10-30mAcm$^{-2,}$ slow stirring pulse plating with duty cycles of 80%, 60% and 40% in the frequency of 100Hz/10Hz were used.	Surface-active agent improved the surface morphology of the deposits by increasing the cathodic overpotential and reducing the limiting current density.	Under optimised condition eutectic Sn-Ag bumps of 15µm height at 30µm pitches could be obtained	[50]
2	Acidic	H_2SO_4 based, $SnSO_4$, $AgNO_3$,thiourea, Polyethlene glycol (PEG)-3000	Temperature: 20°C Rotation speed:11-68 rotations/sec	Near eutectic alloy obtained. Deposited films are compact and have low porosity.	High stability due to very low pH. PEG functioned as an additive	[51]
3	Acidic	alkanesulfonic acid, alkanolsulfonic acid,benzenesulfonic acid or similar groups, brighteners from the group of nonionic surfactants or amines or ketones or aromatic aldehydes, antioxidants for tin mainly hydroxyphenyl compound like phenol, catechol, pyrogallol, hydroquinone, ascorbic acid, sorbitol.	pH<2 Temperature:25°C	Films composition: 20-99%Sn/80-1%Ag Film thickness:1-3µm	Stable for a long time (beyond 120 hours) even at elevated temperatures	[52]
4	Acidic	sulfuric acid or phophonic acid, alkanesulfonic acid or hydroxycarboxylic acid. thiourea as chelating agent, nonionic surface active agent, some additives from mercapto group. metal salts.	pH<1 , Temperature: 15-35°C, Agitation:cathode rocking/ pumping Current Density: 1-1000mAcm^{-2}	Homogeneous film with good external appearance Film Composition: 99.9-10wt%Sn &0.1-90wt%Ag	Preferential deposition of Ag was eliminated.	[53]

Table 5. (continued)

S no	Bath type	Bath constituents	Process condition	Bath/Film properties	Remarks	Reference
5	Alkaline	$K_4P_2O_7$, $Sn_2P_2O_7$, AgI , KI, PEG of different molecular weights were used as additive	Deposition mode : pulse plating Temperature: 25°C Current density: a periodic current swing of 1 s at 1 mAcm^{-2} and 5 s of current density controlled between 4.5 mAcm^{-2} and 9.0 mAcm^{-2}. Time: 900 s Agitation of electrolyte: yes	Sn-3.5 wt.% Ag film was confirmed with the pulse plating mode at positive current density of 6.0 mAcm^{-2} and negative current density of 1.0 mAcm^{-2} for processing time of 15 min	PEG helped in stabilizing the bath and also functioned as an inhibitive agent	[54]
6	Alkaline	$K_4P_2O_7$, $Sn_2P_2O_7$, AgI , KI . Polyethylene glycol (PEG) – 600 and formaldehyde (HCHO) as additive	Temperature: Room temperature, Current density: 2-20mAcm^{-2}	Near eutectic composition, PEG and formaldehyde together produce films with smooth morphology and very fine microstructure.	PEG inhibited Sn deposition. This inhibition was weakened by HCHO	[55]

Silver-thiourea complex was found to be effective in MSA based electrolyte systems. Cyclic voltametry studies done by this group have shown a clear shift in deposition potential of Ag^+ ions with the addition of thiourea to silver containing MSA electrolytes [49]. This helps in controlling the precipitation of Ag ions in the presence and oxidation of stannous ions to its stannic state. There are reports of sulfur containing organic chelating agents in baths depositing Sn-Ag eutectic films [56]. FTIR and Raman spectroscopic analyses reports confirm silver-thiourea chelation through sulfur-metal bonding [45, 57].

There are a few reports on pyrophosphate based alkaline Sn-Ag deposition baths [54-55]. In these systems, Polyethylene glycol (PEG) as an additive is seen to be effective in inhibiting the deposition of metal ions and, thus, helps in refining the morphology of the deposited film. Konda *et al* report an electroplating bath containing an organic sulfonate bath containing pyrophosphate, iodide and triethanolamine as chelating agents for Sn-Ag deposition. An amine-aldehyde reaction product served as a brightener [58]. If issues in lithographic compatibility can be taken care of, alkaline baths are promising enough to deliver good quality films in the desired composition range and the chemical aggressiveness of the high acidic baths can be done with.

It is known that one of the issues in Sn–Ag deposition is the need of relatively lower concentration of Ag in the eutectic composition. It is reported by Bioh Kim and his group [59] that the deposition parameters have a profound effect on the deposited film. They studied the effect of current density on the microstructure and composition of the deposited film. It was concluded that dendrites result from the mass transfer limitation of silver ions. Tin deposits preferentially with increasing current density, which suppresses silver dendrites and develops facets . The development of facets results from the preferred orientation (or texture) of tin growth. Under mass transfer control of tin, the morphology changes again to dendrites.

It is reported that the alloy composition was also sensitive to other conditions, such as, silver concentration, bath temperature, flow conditions, pattern parameters for bumping process *etc* [59]. The limiting Ag concentration at high current density is a useful result, but there is deterioration of microstructure with current density. Further work in this direction may be necessary.

4.4. Tin–Silver–Copper

The relatively lower melting temperature and better mechanical properties compared to Sn-Ag and Sn-Cu alloys has put Sn-Ag-Cu alloy on the forefront as a potential lead-free candidate. Researchers are exploring the possibility of optimizing a bath that could co-deposit the ternary eutectic or at least near eutectic alloy combination. To the best of our knowledge, there are only a handful of reports on the co-deposition of Sn-Ag-Cu, possibly due to the inherent difficulties already discussed elsewhere. A gist of the available reports on SnAg-Ch baths is presented in Table 6.

An alkaline bath with a stability of over six months is reported which uses two different chelating agents. The pH of the solution was varied in the range from 7.5 to 9.5 and the deposition was carried out at temperatures varying from 22°C to 50°C under various stirring rates. It was found that the pH did not have any effect on the film properties but the films became increasingly dendritic as the temperature and the stirring rates were increased at constant current density [63]. Amongst the acidic baths, two types of electrolytes have been reported for these ternary depositions, *viz.* sulfuric acid and MSA. The sulfuric acid based bath uses thiourea as chelating agent and POELE as a surfactant [62]. It was found during the investigations that thiourea, apart from being a chelating agent, also gets adsorbed on the deposited Cu, Ag and Sn and inhibits further deposition. POELE, at the same time, was found to adsorb and inhibit only Sn deposition. Thiourea and POELE both together helped in obtaining a smooth eutectic Sn3.5Ag0.75Cu film. Although sulfuric acid baths containing thiourea also gives favorable results, the corrosive nature of sulfuric acid raises concern.

There is an important distinction between the two types of MSA based baths reported for the deposition of Sn-Ag-Cu films. The one reported by Jinqiu Zhang and his group [60, 61], uses three different chelating ions for the three metal ions simultaneously in the bath and successfully deposited films in the desired composition range.

While potassium pyrophosphate chelates Sn, KI and triethanolamine (TEA) were added in the bath as a chelating agent for Ag and Cu respectively.

Table 6. Summary of reported plating baths for the co-deposition of binary Sn-Ag-Cu films

S no	Bath type	Bath constituents	Plating condition	Bath/Film properties	Remarks	Reference
1	Acidic	MSA, SnMS, AgI, CuMS $K_4P_2O_7$,KI,HQ,TEA, Heliotropin	PH:5.5, Temperature : $20^{\circ}C$ Current Density: 30-50mAcm^{-2} Agitation: nil	Film contains 88-95wt%Sn-5-10%Ag-0.5-2%Cu. bright, compact and smooth deposits. HT is the main brightening agent and increases the cathodic polarization and refines the grains, TEA is a brightening promoter that reduces the cathodic polarization and densifies the deposits. Sn forms transient complexes withTEA	$K_4P_2O_7$ is chelating agent for Sn TEA is a chelating agent for Cu. HQ is antioxidant. KI is a chelating agent for Ag	[60], [61]
2	Acidic	H_2SO_4, SnSO$_4$,CuSO$_4$,AgNO$_3$, thiourea, POELE	Temperature: $20^{\circ}C$, Current Density: 20mAcm^{-2}	Thiourea chelates Ag^+ and Cu^{2+}, POELE increases the smoothness of the deposited film thiourea adsorbs on Sn, Ag & Cu but POELE adsorbs only on Sn and controls its deposition. Film composition: Sn-3.5Ag-0.75Cu		[62]
3	Acidic	H_2SO_4 based, SnSO$_4$, AgNO$_3$, thiourea as chelating agent, PEG-3000 as additive	Temperature: $20^{\circ}C$, Rotation speed:11-68rotations/sec	Near eutectic Sn-Ag-Cu film. Deposited films are compact and have low porosity	High stability due to very low pH	[51]
4	Acidic	MSA, Sn methane sulfonate, Ag$_2$(SO4)$_3$, CuSO$_4$, thiourea, iso-octyl phenoxy polyethoxy ethanol (OPPE)	Temperature: $20^{\circ}C$, Current Density:2-20 mAcm^{-2} PH: ~4	Near eutectic Sn-Ag-Cu composition. OPPE helps in improving the bath stability and also in refining the film microstructure. Thiourea effectively chelates Cu^{2+} & Ag^+	Stability - about a week	[49]
5	Alkaline	Cyanide-free bath with Sn, Ag,and Cu salts, two chelating agents	Temperature : 22,30,40 & 50°C pH: 7.5,8.5 & 9.5, Agitation: 0,100 and 200 rpm,	Near eutectic Sn-Ag-Cu composition.	Stability: More than Six months	[63]

An additive, Heliotropin (HT) also contributes to the deposition mechanism. An interesting result reported by these authors indicates that if the cathode polarization caused by individual chelates is balanced and close to the polarization of the bath without additives, the films that are produced have improved microstructure. Figure 6 reproduces the linear Sweep voltametry (LSV) curves for the bath with and without TEA and HT.

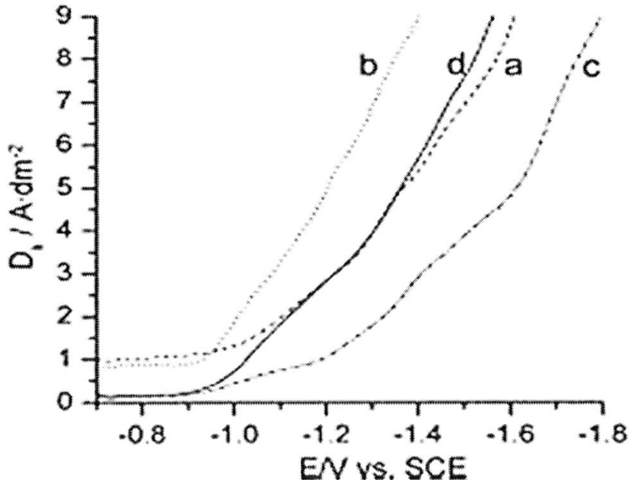

Figure 6. Linear Sweep Voltametry curves of various MSA based baths for Sn-Ag-Cu deposition: (a) basic bath, (b) basic bath + 0.225 mol/L TEA, (c) basic bath + 6.4 mol/L HT and (d) basic bath + 0.225 mol/LTEA + 6.4 mmol/L HT [61, Reprinted from *Electrochim Acta*, 53, Jinqiu Zhang; Maozhong An; Limin Chang; Guiyuan Liu, 'Effect of triethanolamine and heliotropin on cathodic polarization of weakly acidic baths and properties of Sn–Ag–Cu alloy electrodeposits',2637-2643, Copyright (2008), with permission from Elsevier]

It can be observed from Figure 6 that TEA decreases the cathodic polarization of the bath while HT increases the same. When both these chemicals are present (curve d) in the bath, they are balanced and the polarization is somewhere between the two. Figure 7 presents the microstructure of the films deposited under condition 'a' to 'd' in Figure 6. The favorable effect of both the additives *viz.* HT and TEA on the microstructure can be easily observed. It may be noted that HT is the main brightener and TEA acts as a brightener promoter. This bath produced films with composition in the range of 88-95wt%Sn-5-10%Ag-0.5-2%Cu. The silver content in the film is little on the higher side and needs to be improved. Further, the large number of additives in this bath is a cause of concern.

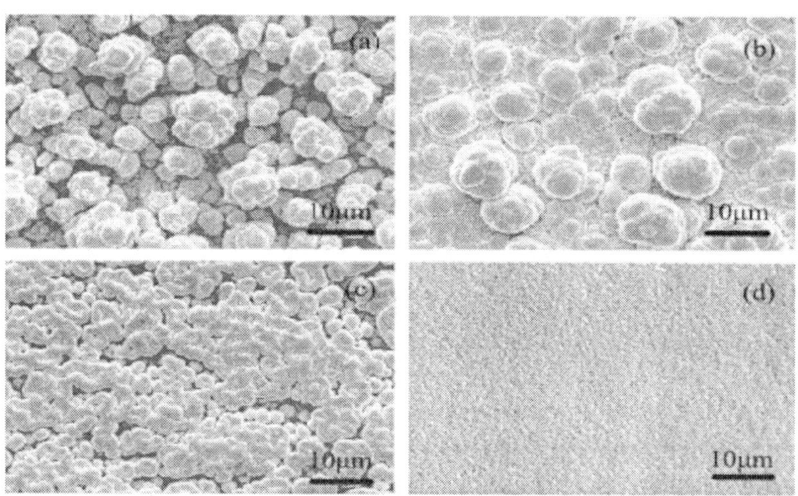

Figure 7. SEM images of Sn–Ag–Cu alloy electrodeposited from various baths at 4 A/dm2 for 5 min on Cu plates: (a) basic bath, (b) basic bath + 0.225 mol/L TEA, (c) basic bath + 6.4 mmol/L HT and (d) basic bath + 0.225 mol/L TEA + 6.4 mmol/L HT [61, Reprinted from *Electrochim Acta*, 53, Jinqiu Zhang; Maozhong An; Limin Chang; Guiyuan Liu, 'Effect of triethanolamine and heliotropin on cathodic polarization of weakly acidic baths and properties of Sn–Ag–Cu alloy electrodeposits',2637-2643, Copyright (2008), with permission from Elsevier]

Using thiourea that chelates both Ag^+ and Cu^{2+} is one option that may reduce the number of bath components. Here an optimum amount of thiourea needs to be determined to keep both the electropositive metal ions in bonding.

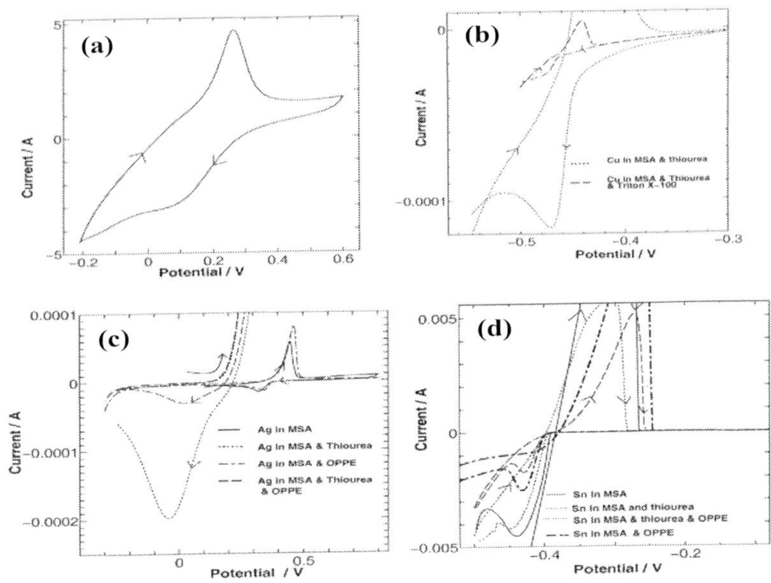

Figure 8. Cyclic voltametric curves for different baths. a) Cu^{2+} in MSA, b) Cu^{2+} in MSA with and without thiourea and OPPE c) Ag^+ and d) Sn^{2+} in individual bath formulations with and without Thiourea and OPPE[49, Reprinted from *Surf. Coat. Tech*, 202, Shany Joseph; Girish J Phatak, 'Effect of Surfactant on the bath Stability and electrodeposition of Sn-Ag-Cu films', 3023, Copyright (2008), with permission from Elsevier]

A surplus amount may introduce sulphur into the films to form metal sulphide and insufficient amount may cause metal precipitation during the Sn^{2+} oxidation to Sn^{4+} state. Our report on MSA based co-deposition bath [49] reports a formulation containing only thiourea as the chelating agent and small amount of a surfactant, OPPE (Triton X-100®) that gives a stable bath and deposits a near eutectic Sn-Ag-Cu film. The presence of surfactant refined the microstructure of the deposited films and helped in improving the bath stability. A detailed study undertaken to ascertain the nature of bonding of OPPE with the metal ions/chelates with thiourea, indicate that OPPE may only be forming some loose hydrogen bond that alters the reduction potential marginally. On the other hand, clear evidence of large shift in reduction potentials could be seen for Ag and

Cu ions with thiourea. This can be observed from Figure 8 that presents the cyclic voltametric curves of the three metals in the presence and absence of thiourea and OPPE. Further, the cyclic voltametric curve of the ternary bath presented in Figure 9 show a single broad peak essentially indicating that the individual metal depositions are taking place in a narrow range of potential. Such a trend highlights the effective chelation by the chelating agent, altering the deposition potential of the metal ions. This would ensure better reliability of the deposited films. It is found that the film composition is independent of current density beyond $5\,mAcm^{-2}$ which is evident from Figure 10. From the bath utility point of view, this is a useful result.

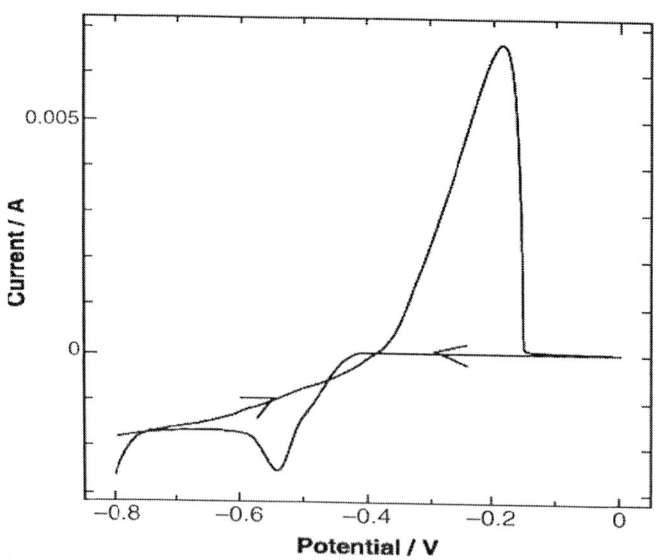

Figure 9. Cyclic voltametry of ternary Sn-Ag-Cu bath [49, Reprinted from *Surf. Coat. Tech*, 202, Shany Joseph; Girish J Phatak, 'Effect of Surfactant on the bath Stability and electrodeposition of Sn-Ag-Cu films', 3023, Copyright (2008), with permission from Elsevier]

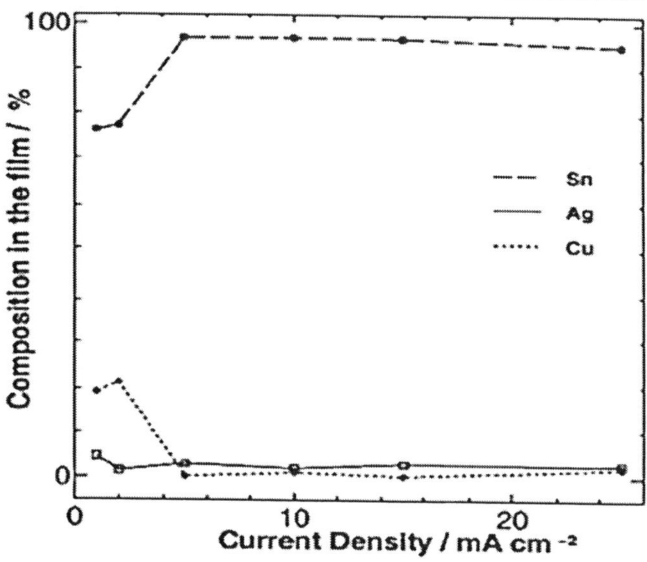

Figure 10. Effect of current density on the elemental composition of the deposited Sn-Ag-Cu films [49, Reprinted from *Surf. Coat. Tech*, 202, Shany Joseph; Girish J Phatak, 'Effect of Surfactant on the bath Stability and electrodeposition of Sn-Ag-Cu films', 3023, Copyright (2008), with permission from Elsevier]

The said bath was subjected to impedance analysis at varying thiourea concentrations named TU0-TU4 for thiourea concentrations 0.06M, 0.13M, 0.16M and 0.26M respectively. Figure 11 presents the impedance spectra of baths in high and medium frequency range at different thiourea concentrations. It concluded from the analysis of these curves and similar recordings for different baths prepared by elimination of metal ions and additives, that thiourea may be responsible for the broadened reduction peak of the ternary bath (Figure 9) [64]. Higher concentrations of thiourea seem to be helping in obtaining refined microstructure. This can be observed from Figure 12 that presents the SEM micrographs of films deposited with various concentrations of thiourea. The above impedance analyses also revealed that there is a formation of an inhibiting film on the cathode surface causing negative real impedance at high frequency and that the metal ions deposit competitively against the removal of this inhibiting film. It is concluded that this film is contributed either by MSA or water molecules [64]. These results throw new light on the behavior of an important chelating agent found in thiourea, and, for the first time, present impedance analysis of Sn-Ag-Cu ternary bath.

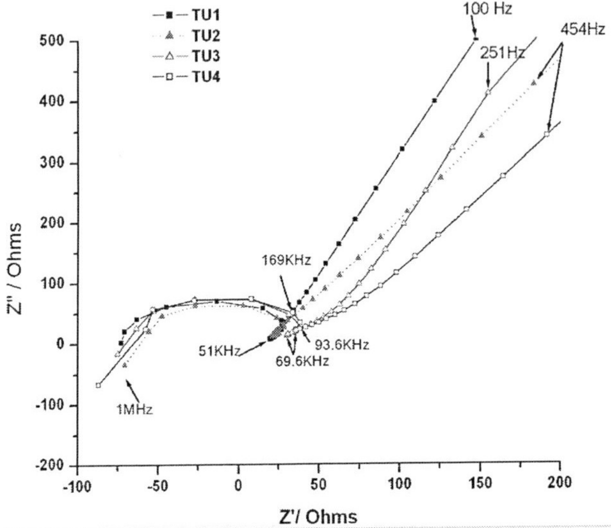

Figure 11. Nyquist plots for ternary Sn-Ag-Cu baths with compositions TU1-TU4, in high frequency range [64]

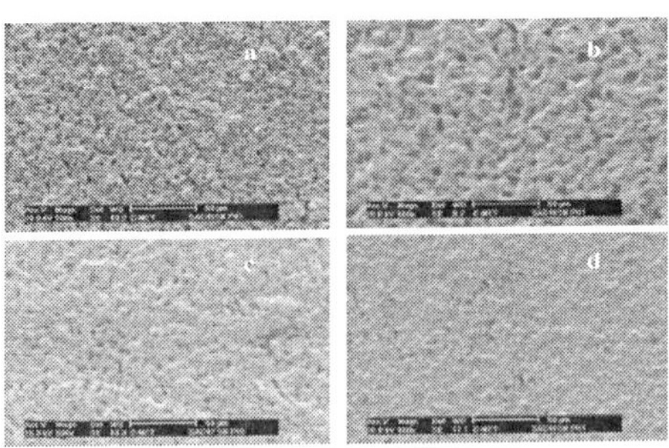

Figure 12. SEM Microstructure of ternary Sn-Ag-Cu films deposited thiourea concentrations a) TU1 b) TU2 c) TU3 d) TU4 [64]

4.5. TIN-BISMUTH

Normally, in the Sn-Bi co-deposition bath, the 'noble' Bi starts depo-siting at lower potentials and it becomes difficult to control its composition. Thus, additives are necessary. Amongst the different additives explored to control the bath and film properties is POELE, reported by Fukuda and his group [65], for a sulfuric acid based bath. POELE gets adsorbed.on the deposited Bi and Sn and thus help in obtaining a smooth and uniform microstructure. Ethylene diamine tetra acetic acid (EDTA) forms stable complexes with Bi in the pH range 1-3. However, when used in Sn-Bi co-deposition bath, the complexes were not very stable, especially when the pH was raised beyond 2. Addition of polycarboxylic group compounds, such as gluconic acid and glutamic did help in improving the stability [66]. Similar results were reported for Sn-Bi depositions using stannic chloride solutions where a combination of EDTA, citric acid and PEG had to be used for attaining the bath stability as well as uniform and smooth film [67]. MSA baths for Sn-Bi eutectic depositions with commercial additives and grain refiners are reported [68, 69]. Some patented information on Sn-Bi [70, 71] baths containing polycarboxylic group compounds are reported. These baths are highly acidic, show good stability and deposit films with no whiskers when tested using accelerated conditions for one week. These baths deposit Sn-Bi films with Bi composition varying from 0.1 to 75% over a wide current density range.

4.6. TIN-BISMUTH-COPPER:

The Sn-Bi-Cu ternary system is not reported for its direct use as a lead-free solder alloy. However, this bath was found useful for obtaining solder deposition of quaternary Sn-Ag-Bi-Cu system due to the poor stability of its quaternary co-deposition bath. The only alternative to this difficulty was to use a combination of sequential and co-deposition techniques. Thus, the quaternary system could be obtained by first electroplating silver, followed by co-deposition of Sn-Bi-Cu [72].

The ternary Sn-Bi-Cu bath based on sulfuric Acid and primarily containing stannous ions (Sn^{2+}) obtained from its sulphate salt, has been reported [66]. The Bi and Cu ions were also contributed by their respective sulphate salts, while EDTA and Polycarboxylic acid group compound were used as chelating agent and surface finishing agents respectively, which helped in controlling the surface morphology. The pH of the bath was adjusted to ~4.5 using acetate buffers. The bath showed good stability for more than a month and the film properties showed good consistency. The film deposited after 72 hours of bath preparation gave compact films with composition similar to the fresh bath deposits. Figure 13 presents the composition of Sn-Bi-Cu films obtained after varying hours of bath preparation which indicates the stability and consistency of the bath. It was found that in the current density range of 5-25mAcm^{-2} the deposited films have the same composition range. Figure 14 presents the dependence of film composition on current density, which indicates that the composition is constant over the current density range of 1-25mAcm^{-2} [66]

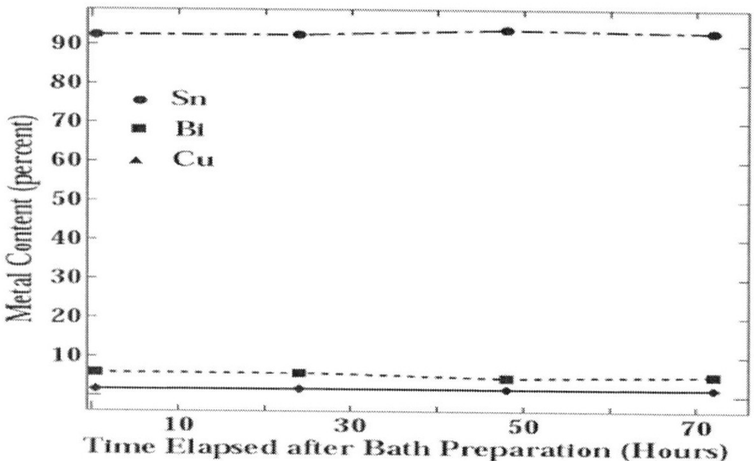

Figure 13. Composition of the deposited Sn-BiCu films deposited after hours of bath preparation [66]

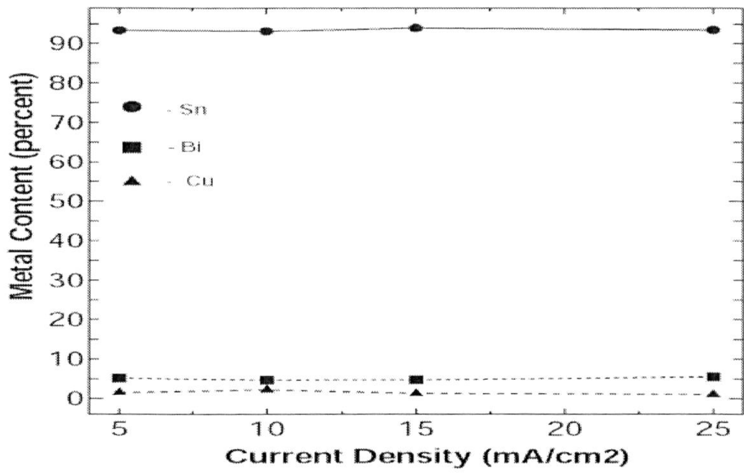

Figure 14. Effect of current density on the composition of the Sn-Bi-Cu films [66]

Another, similar stannic (Sn^{4+}) ion based Sn-Bi-Cu bath with excellent stability is reported [73]. Sulphate salts of Bi and Cu were used while stannic sulphate was precipitated from stannic chloride. In this bath as well, EDTA and polymonocarboxylic acid served as chelating agent and surface finishing agent, respectively. The cyclic voltametry curve of this bath is presented in Figure 17. It can be clearly observed that there is a single broad deposition peak between 0.04 V and -0.15 V with no other discernible peaks. Existence of a single peak is an indication of co-deposition of the three metal ions. Figure 18 shows the effect of current density on the composition of the deposited film. It is observed that the films had higher noble metal content at lower current density ($<10mAcm^{-2}$) which reduced with increasing current density up to $10mAcm^{-2}$. The three metals reach their limiting current density at $10mAcm^{-2}$ and there is no change in composition with further increase in current density. The bath which deposited films close to eutectic compositions of Sn–Bi with <5 wt. % Cu content had very good stability and did not show any precipitation even beyond two months [73]

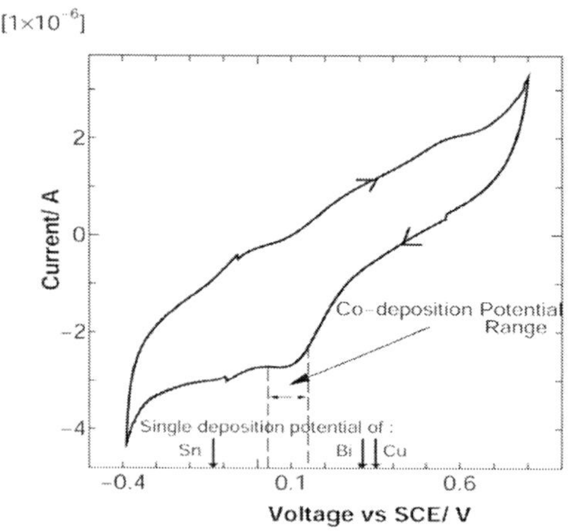

Figure 15. Cyclic voltametry curve for Sn–Bi–Cu bath with Sn^{4+} ions [73, Reprinted from *Journal of Applied Electrochemistry*, 36, Shany Joseph; Girish, J; Phatak, K. Gurunathan, Tanay Seth, D.P. Amalnerkar And T.R.N. Kutty,, Electrochemical co-deposition of ternary Sn-Bi-Cu films for solder bumping applications', 907–912, Copyright (2006), with permission from IEEE

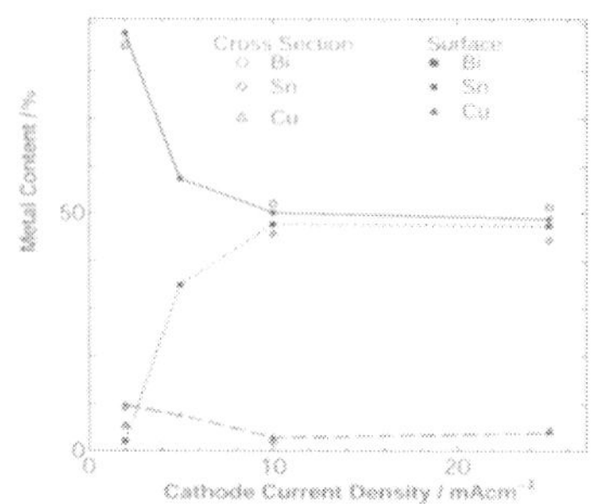

Figure 16. Effect of current density under static bath conditions for Sn-Bi-Cu films with Sn^{4+} ions [73, Reprinted from Journal of Applied Electrochemistry, 36, Shany Joseph; Girish, J; Phatak, K. Gurunathan, Tanay Seth, D.P. Amalnerkar And T.R.N. Kutty,, Electrochemical co-deposition of ternary Sn-Bi-Cu films for solder bumping applications', 907–912, Copyright (2006), with permission from IEEE]

4.7. TIN-ZINC

The Sn-Zn is another alloy with comparatively lower melting point and good mechanical properties. However, its poor wetting properties are its drawback. The eutectic composition of Sn-9Zn has a melting point of $193°C$. Most of the plating baths reported for Sn-Zn are acidic, and tartaric acid and calcium gluconate seem to be the most commonly used chelating agents for this alloy combination. Sulfate baths containing Sulfate salts of Sn and Zn have only sodium sulphate as an additive and tartaric acid as chelating agent. Both the metal ions are chelated by tartarate ion and the films, under optimized plating condition, have smooth and uniform morphology with round grains [74, 75]. The bath can be used in mild acidic conditions at room temperature. Sodium gluconate is another chelating agent used for Sn-Zn codeposition using sulphate baths at pH of about 4 [76]. The morphology of the deposited films is fine, regular, dense and compact for films with higher Zn content. As Sn content in the bath starts increasing, this morphology tends to disappear and a crystallite coating, higher in Sn, takes its place. Sn as stannic chloride along with $ZnCl_2$ can be used to formulate a chloride based bath with calcium gluconate as chelating agent. Good deposit quality at close to neutral pH is the highlight of this bath. The deposit quality can be refined by varying Sn composition in the electrolyte [77].

Chapter 5

Conclusion

The use of solders in electronics is decades old, and it is expected to continue for a conceivable future. The form and composition of the solder, however, has been undergoing changes due to changes in soldering processes. The recently imposed change on the solders is rather external to the electronics, caused mainly due to the environmental and health hazards emanating out of lead used in the solders. The search for a suitable replacement to Sn-Pb solders has been quite illusive. It is seen that amongst various lead-free options, the binary, ternary and some of the quaternary alloys of Ag, Cu, Bi, Zn and In with Sn have properties closer to the Sn-Pb benchmark. One of the important and relatively recent applications of solders in electronics is solder bumps that provide a high density interconnection alternative to wire bonding to the Si chips and through-hole pins on the packages. These solder bumps are obtained through a reflow of deposited solder films on wetting pads. It is seen that co-deposition of the lead-free elements in required composition by electroplating is amongst the most advantageous processes to obtain these solder films.

Electroplating of binary and ternary combinations of lead-free solders containing Sn is always challenging due to peculiarly different reduction potential of Sn in comparison with the co-deposition elements, *viz.* Cu, Ag, Bi, Zn and In. This calls for use of chelating agents and special additives in the plating bath, and concerted efforts in formulating them precisely to obtain optimum results. There is a range of binary and ternary plating baths reported for the deposition of these solders. It is observed that there has been a considerable effort to obtain ideal plating bath to co-deposit binary and ternary compositions of the available lead-free options. Increasingly, the efforts are directed towards the development of environmentally friendly plating baths and processes with lower harshness, even though some of the traditional and commercially successful options were available for direct use. Thus, there is relatively low interest in the commonly available cyanide based baths.

There are two major options in the bath systems, *viz.* acidic and alkaline baths. It is observed that the constitution of an alkaline bath is much simpler as it seldom suffers from the disadvantage of having to deal with the two oxidation states of Sn simultaneously. Additives are still necessary, but those are limited to surface active agents, used mainly as brightener, leveler leading to refinement of the film microstructure, although a very few reports indicate use of chelating agents. Due to the simplicity of the baths the alkaline baths are attractive options. However, for the emerging lead-free solder applications in electronics that necessarily require photolithographic patterning prior to solder deposition, it is difficult to opt for alkaline baths due to instability of the cross-linked photo-polymer in this pH range. Additionally, the reports indicate that these baths do not support high current density, limiting them to only moderate growth rates.

The Sn-Cu binary plating in acidic baths can be carried out using sulfuric acid, MSA or chloride based systems. Many chelating agents and additives have been explored. Thiourea for MSA and sulfuric acid baths, while triammonium citrate for chloride baths has been reported. Thiourea-copper complexes are quite stable and have been used many applications in electroplating, biological applications and many others. This compound along with additives, such as, tartarate, POELE, OPPE, triethanolamine *etc* can help in formulating an electroplating bath for this binary alloy with desired properties. The results bring out the role of surfactants such as Laprol and POELE in helping co-deposition of Sn-Cu and obtaining stable bath in absence of any specific chelating agent. Similar to Sn-Cu, the Sn-Ag binary plating baths also seem to be using Sulfuric acid and MSA as base electrolytes and thiourea as chelating agent. The thiourea-silver complexes have been reportedly effective in Sn-Ag depositions. The film deposition results for both Sn-Cu and Sn-Ag indicate acceptable microstructure, and films with eutectic or near-eutectic composition. The very fact that thiourea forms chelates with Ag as well as Cu, makes it one of the promising chelating agents for the ternary Sn-Ag-Cu co-deposition baths. Interestingly, separate chelating agents for all the three metal ions using pyrophosphate, iodide and triethanol amine, have also been reported. In comparison, thiourea may be a better option due to the reduced bath complexity. An interesting feature of the MSA based Sn-Ag-Cu bath is the current density independence observed due to the limiting current density, which can be exploited for consistent results. Similar effect is seen for the sulfate baths for depositing Sn-Bi-Cu films. The Sn-Bi depositions are reported using chloride ions, sulfuric acid and MSA based electrolytes Chelating agents, such as, EDTA and citric acid for chloride, thiourea and polycarboxylic acid for sulfuric acid and some commercial compounds for MSA electrolytes are reported. The Zn-tartarate complexes for sulphate and Zn-gluconate chelates for chloride based electrolytes are reported to be effective in Sn-Zn depositions. The effect of additives, however, needs to be explored further.

From the above discussion it is seen that there are primarily three options in acidic baths, *viz.* sulfuric acid based baths, chloride based baths, and methane sulphonic acid based baths, all providing a varying degree of success. The difficulty in these baths is caused by the oxidation process to form Sn^{4+} ions, which reveals itself in the form of bath instability, and, additionally reduces the number of ions of Sn participating in the deposition process at a given potential. This leads to the variation in the composition of the film. These baths, therefore, essentially require anti-oxidants and chelating agents to obtain any practical bath stability. Amongst the three, it is seen that the MSA baths have shown better results for both, binary and ternary bath compositions. The MSA based baths have an advantage of being less corrosive due to medium pH range, which is an advantage. These baths are environmentally friendly, as these are much safer to work with and the effluent treatment is comparatively cheap. Most of the metal salts are highly soluble in MSA and this bath is reported to take higher current densities even up to $2000 mAcm^{-2}$. Amongst the chelating agent, thiourea as a common chelating agent for both, Ag and Cu and seems useful. Addition of thiourea derivatives prevents whiskers from being deposited and allows uniform and sufficiently smooth plating [78].

It is seen that most of the co-deposition efforts to obtain lead-free solder depositions are far from complete. Complete bath characterization, studies about the effect of each additive on deposited films and other plating conditions are still underway. These studies would eventually lead to the optimization of bath composition and plating conditions, when bump formation and related studies can be taken up. Hopefully, along the way, the bath related factors as well as the solder applications in electronics would help in arriving at an appropriate lead-free alloy composition that is acceptable under all conditions. As of today, the Sn-Ag-Cu lead-free composition and its MSA based baths look favorites.

ACKNOWLEDGMENTS

The authors wish to thank Prof. T. R. N. Kutty, Materials Research Centre, Indian Institute of Science, Bangalore, India, for guidance and support during the reported research activity. We are also thankful to Dr. D. P. Amalnerkar, Executive Director, Centre for Materials for Electronics Technology (C-MET), for his unflinching support. This work was fully supported by C-MET through the generous grants from the Department of Information Technology, Government of India. Their contributions are gratefully acknowledged.

REFERENCES

[1] Klaus Zwilsky, M; Edward Langer, L. *"Electronic materials Handbook-* Volume I: Packaging, ASM International, 1989.

[2] Schlesinger, Mordechay; Paunovic, Milan. *"Modern electro plating"* (Wiley-Interscience Publication) 2000.

[3] Puttlitz, Karl J; Stalter, Kathleen A. *Handbook of Lead-free Solder Technology for Microelectronic Assemblies*, CRC Press, New York, 2004.

[4] Chun-Jen Chen and Kwang-Lung Lin, 'The reactions between electroless Ni-Cu-P Deposit and 63Sn-37Pb flip chip solder bumps during reflow', *J. Electron. Mater*, 2000, 29, 1007.

[5] Lau, JH; Chang, C; Lee, SWR. 'Failure analysis of solder bumped flip chip on low-cost substrates' *IEEE Trans. Electron. Packg. Manufact.*, 2000, 23, 19.

[6] Kwang-Lung Lin; Yi-Cheng Liu, "Reflow and property of Al/Cu/ electroless nickel/Sn-Pb solder bumps", *IEEE trans adv packag*, 1999, 22, 568.

[7] Kwang-Lung Lin; Yi-Cheng Liu, "Manufacturing of Cu/electroless nickel/Sn-Pb flip chip solder bumps" *IEEE trans adv packag*, 1999, 22, 575.

[8] Rao R. Tummala, Eugene J. Rymaszewski and Alan G. Klopfenstein, *"Microelectronics Packaging Handbook-* vol – II" (Kluwer Academic Publishers, New York) 1999.

[9] Thorsten Teutsch; Ronald G. Blankenhorn; Elke Zakel, "Lead-Free Solder Bumping Process for High Temperature Automotive Application" *proceedings of ECTC*, 2003, 1468- 1471

[10] Klaus, Ruhmer; Eric, Laine; Karin, Hauck; Dionysios, Manessis; Andreas, Ostmann; Michael, Toepper, 'Low Cost UBM for Lead Free Solder Bumping with C4NP', *Proc. of HDP*, 2007.

[11] Glenn, R. Blackwell. *"The Electronic Packaging Handbook"* (A CRC press & IEEE press) 1999.

[12] Lau, JH. *"Flip Chip Technologies"* (Mc Graw-Hill, New York) 1996.

[13] Li Li & Pat Thompson, 'Stencil Printing Process Development for flip chip interconnect', *"IEEE Trans. Electron. Packag. Man ufact."* 2000, 23, 165.

[14] Tan, AC. *"Tin solder plating in semiconductor industry"* (Chapman & Hall publication) 1993.

[15] Rao R. Tummala; Eugene J. Ryma; Zewski, S; Alan G. Klopfenstein, *"Microelectronic Packaging Handbook-* part III" (Chapman & Hall Publication) 1993.

[16] Mulugeta Ablew & Dr. Guna Selvaderey, 'Lead-free solders for surface mount technology applications (part1)', *"Chip Scale Review"*, September 1998.

[17] Price, JW. *"Tin and tin alloy plating"* (Electrochemical publications limited) 1983.

[18] John H. Lau; Ricky Lee, SW. *"Microvias-for low cost high density interconnects"* (Mc Graw Hill publication) 2001.

[19] Harrison, MR; Vincent, JH; Steen, HA. 'Lead free reflow soldering for electronic assembly', *"Soldering & surface mount Technology"*, 2001, 13, 21.

[20] Seeling, Karl and Suraski, David 'The status of lead-free solder Alloys', Proc. Of *Electronic Components and Technology Conference*, (2000) 1405-9.

[21] Dr. Jennie S. Hwang, 'Bismuth in Electronic Solder (part1)', *"The bulletin of bismuth institute"*.

[22] Dr. Thomas Siewert, Dr. Stephen Liu, Dr. David R. Smith, Mr. Juan Carlos Madeni, *'Database for Solder Properties with Emphasis on New Lead-free Solders: Properties of Lead-Free Solders* '(Release 4.0), National Institute of Standards and Technology, & Colorado School of Mines http://www.boulder.nis t.gov/div853/lead_free/props01.doc Accessed on 18-06-09

[23] Low, CTJ; Walsh, FC. 'Electrodeposition of tin, copper and tin– copper alloys from a methanesulfonic acid electrolyte containing a perfluorinated cationic surfactant', *Surf. And Coat. Tech,* 202 (2008) 1339–1349

[24] Martyak, Nicholas M & Seefeldt, Robert 'Additive effects during plating in acid tin methane sulfonate electrolytes' *Electrochimica Acta*, 2004, 49, 4303-4311

[25] Low, CTJ; Walsh, FC. The stability of an acidic tin methanesulfonate electrolyte in the presence of a hydroquinone antioxidant *Electrochimica Acta* 53 (2008) 5280–5286

[26] Carlos, IA; Bidoia, ED; Pallone, EMJA; Almeida, MRH; Souza, CAC. "Effect of tartrate content on aging and deposition condition of copper–tin electrodeposits from a non-cyanide acid bath:, *Surf. And Coat. Tech.*, 2002, 157, 14-18.

[27] Chunfen, Han; Qi, Liu & Douuglas, G Ivey. 'Development of simple Electrolytes for the Electrodeposition and Electrophoretic Deposition of Pb-free, Sn-based Alloy Solder Films', *Proc. of CS Mantech conf*, May 14-17 2007, Austin Texas,

[28] Chunfen, Han; Qi, Liu and Douglas, G Ivey, 'Development of simple Electrolytes for the Electrodeposition of Pb-free, Sn-based Alloy Solder Films', *Proc. of Mater. Res.Soc. Symp*, 2007, 993, E04-07

[29] Fukuda, M; Hirakawa, K; Ymatsumoto. *J. Surface Finishing Soc. Jpn.* 50(1999)1125

[30] Bestetti, M; Vicenzo, A and Cavallotti, PL; Politecnico di Milano, 'Tin Copper alloys electroplating from thiourea solutions', *Proc. Of 203rd ECS meetings*,

[31] Seok Won Jung, Jae Pil Jung & Y (Norman) Zhou, 'Characteristics of Sn-Cu solder bump formed by electroplating for Flip Chip', *IEEE Trans. Elec Pckg Mnfg*, 29, No.1 (2006) 10-16

[32] Jeffrey Chang-Bing Lee; Yung-Ling Yao; Fang-Yi Chiang; Zheng, PJ; Liao, CC; Chou, YS. 'Characterization study of Lead-free Sn-Cu Plated Packages', *Proc. ECTC Conf.*, 2002, 1238-1245

[33] Survila, A; Mockus, Z; 'Current oscillations and a negative impedance observed during copper and tin codeposition from solutions involving Laprol 2402C as a surface-active substance' *Electrochim.Acta,* 1999, 44, 1707-1712

[34] Finazzi, GA; De Oliveira, EM; Carlos, IA. 'Development of a Sorbitol alkaline Cu-Sn plating bath and chemical, physical and morphological charecterization of Cu-Sn films', *Surf. Coat. Tech.*, 2004, 187, 377-387

[35] Kaneko; Mitsuru Hatta; Asao Kunii; Mitsuharu. 'Cyanide-free pyrophosphoric acid bath for use in copper-tin alloy plating', *US Patent 6416571*, (July 2002)

[36] Guausand, E; Torrent-Burgués, J. 'Voltammetric Study of Sn(II) Reduction on a Glassy-CarbonElectrode from Sulfate–Tartrate Baths', *Russian Journal of Electrochemistry,* 2006, Vol. 42, No. 2, 141–146.

[37] Chunfen Han*; Qi Liu; Douglas G. Ivey, 'Kinetics of Sn electrodeposition from Sn(II)–citrate solutions', *Electrochimica Acta,* 2008, 53, 8332–8340

[38] Louis Meites. 'Polarographic Studies of Metal Complexes. I. The Copper(I1) Tartrates', *Journal of the American chemical society,* 71 No.10 (1949) 3269-3275

[39] Chunfen, Han; Qi, Liu; Ivey Douglas, G. 'Electrodeposition of Sn-0.7 wt% Cu Eutectic Alloys from Chloride-Citrate Solutions', *Plating and surface f inishing,* 2008, vol. 95, n°8, 21-29 .

[40] Ben-Zhan, Zhu*; William, Antholine‡, E; Balz, Frei. 'Thiourea protects against copper-induced oxidative damage by formation of a redox-inactive thiourea-copper complex', *Free Radical Biology and Medicine,* 2002, 32, No 12, 1333-1338

[41] Gherrou. H; Kerdjoudj, R; Molinari, E. Drioli, 'Effect of thiourea on the facilitated transport of silver and copper using a crown ether as carrier', *Separation and Purification Technology,* 2001, 22-23, 571-581

[42] Swaminathan K; & Irving, HMNH. 'Infra-red Absorption Spectra of Complexes of Thiourea', *J. Inorg.Nucl Chem,* 1964, 26, 1291-1294

[43] Nisit, Tantavichet, Mark D. Pritzker, 'Effect of plating mode, thiourea and chloride on the morphology of copper deposits produced in acidic sulphate solutions', *Electrochim. Acta,* 2005, 50, 1849-1861

[44] Cofre, P; Bustos, A. 'Voltametric behavior of the copper(II)-thiourea system in sulfuric acid medium at platinum and glassy carbon electrodes', *J of Appl. Electrochem,* 1994, 24, 564-568

[45] Fleischmann, M; Hill, IR; Sundholm, G. ' A Raman spectroscopic study of thiourea adsorbed on silver and copper electrodes', *J Electroanal. Chem.* 157(1983) 359-368

[46] Brown, GM; Hope, GA; Schweinsberg, DP; Fredericks, PM. 'SERS study of the interaction of thiourea with a copper electrode in sulfuric acid solution', *J of Electroanal Chem.*, 1995, 380, 161-166

[47] Stankovic, ZD; Vukovic, M. 'The Influence of Thiourea on Kinetic Parameters on the Cathodic and Anodic Reaction at different metals in H_2SO_4 Solution', *Electrochim. Acta*, 1996, 41, No16, 2529-2535

[48] Hwang, H; Hong, SM; Jung, JP; Kang, CS. "Pb-free solder bumping for flip chip package by electroplating," *Soldering Surf. Mount Technol.*, 2003, vol. 15, no. 2, 10–16.

[49] Shany Joseph; Girisg J Phatak, 'Effect of Surfactant on the bath Stability and electrodeposition of Sn-Ag-Cu films', *Surf. Coat. Tech*, 2008, 202, 3023-3028.

[50] Kim, JY; Yu, J; Lee, JH; Lee, TY. 'The effect of Electroplating Parameters on the composition and Morphology of Sn-Ag Solder', *J of Electron. Mater.*, 2004, 33, No.12, 1459-1464

[51] Ozga, P. 'Electrodeposition of SnAg and Sn-Ag-Cu alloys from thiourea aqueous solutions' Archives of metallurgy and materials, 2006, 51, issue 3, 413-421

[52] Oshima; Katsuhide; Yuasa; Satoshi. 'Acid tin-silver alloy electroplating bath and method for electroplating tin-silver alloy', *U.S. Patent 5, 911, 866*, June 15, 1999

[53] Yanada; Isamu; Tsujimoto; Masanobu; 'Tin-silver alloy electroplating bath and tin-silver alloy electroplating process', *U.S. Patent 6,099,713*, August 8 2000

[54] Hsiao- Yun Chen; Chih Chen; Pu-wei Wu; Jia- min Sheih; Shing-Song Cheng; Karl Hensen. 'Effect of polyethylene Glycol Additives on Pulse Electroplating of Sn-Ag solder', *J of Electron. Mater.*, 2008, 37, number 2, 224-230.

[55] Susumu, Arai; Hideki, Akatsuka; Norio, Kaneko. ' Sn-Ag Solder Bump formation for Flip Chip Bonding By Electroplating', *J of electochem Soc.*, 2003, 150 no. 10 C730-C734.

[56] Se-young Jang; Juergen Wolf; Oswin Ehrmann; Heinz Gloor; Herbert Reichl; Kyung-Wook Paik. 'Pb-free Sn/3.5Ag electroplating bumping process and under bump metallization (UBM) ', *IEEE Trans. Electron Packg Manufg.*, 2002, 25, No.3, 193-202

[57] Stella Nunziante Cesaro, 'FTIR study of a silver–thiourea complex generated in argon and nitrogen cryogenic matrices', *Vibrational Spectroscopy*, 1998, 16, No. 1, 55-59

[58] Kondo, T; Obata, K; Takeuchi, T; Masaki, S. 'Bright tin-silver alloy electrodeposition from an organic sulfonate bath containing pyrophosphate, iodide & triethanolamine as chelating agents', *Plating and surface finishing*, 1998, 85 No2, 51-55

[59] Bioh Kim; Charles Sharbono; Tom Ritzdorf; Dan Schmauch. 'Comparison of Near-Eutectic SnPb and SnAg Solder Plating for WLP Application' Presented at Pan Pacific Microelectronics Symposium, Big Island of Hawaii, Jan. 17-19, 2006

[60] Jinqiu Zhang; Maozhong An; Limin Chang. 'Study of the electrochemical deposition of Sn–Ag–Cu alloy by cyclic voltammetry and chronoamperometry', *Electrochim Acta*, 2009, 54, 2883-2889

[61] Jinqiu Zhang; Maozhong An; Limin Chang; Guiyuan Liu. 'Effect of triethanolamine and heliotropin on cathodic polarization of weakly acidic baths and properties of Sn–Ag–Cu alloy electrodeposits ', *Electrochim Acta*, 2008, 53, 2637-2643

[62] Mitsunobo Fakuda; Kohei Imayoshi; Yasumichi Matsumoto. 'Effects of Thiourea and Polyoxyethylene Lauryl Ether on Electrodeposition of Sn-Ag-Cu Alloy as a Pb- free Solder', *J of Electrochem Soc.*, 2002, 149 No.5, C244-C249

[63] Bioh Kim; Tom Ritzdorf. 'Electrochemically deposited Tin-Silver-Copper Ternary Solder Alloys', *J of Electrochem Soc.*, 2003, 150, No.2, C53-C60

[64] Shany Joseph, 'Development of Lead-free solder bumping by electroplating method for Flip Chip and BGA Application', PhD Thesis, 2009

[65] Mitsunobu Fakuda; Kohei Imayoshi; Yasumichi Matsumoto. 'Effect of polyoxyethylenelaurylether on electrodeposition of Pb-free Sn–Bi alloy', *Electrochim Acta*, 47 (2001) 459-464

[66] Shany Joseph, 'Development of Lead-free (Sn-Ag-Bi-Cu) solder bumps by electroplating process', MSc(Engg) Thesis, 2003

[67] Yi-Da, Tsai; Chi-Chang, Hua; Chi-Cheng, Lin. 'Electrodeposition of Sn–Bi lead-free solders: Effects of complex agents on the composition, adhesion, and dendrite formation', *Electrochimica Acta*, 2007, 53, 2040–2047

[68] Min-Suk Suh; Chan-Jin Park; Hyuk-Sang Kwon. 'Effects of plating parameters on alloy composition and microstructure of Sn–Bi electrodeposits from methane sulphonate bath', *Surface & Coatings Technology*, 2006, 200, 3527–3532

[69] Murphy; Timothy, I; Reynolds, Brian R. 'Low temperature tin-bismuth electroplating system', *U.S. Patent 5,227,046*, July 13,1993.

[70] Wilson; Harold, P; 'Bismuth composition, method of electroplating a tin-bismuth alloy and electroplating bath therefor ', *U.S Patent 4,331,518*, May 25, 1982

[71] Sakurai, Hitoshi, Yuasa, Satoshi, 'Sn-Bi alloy-plating bath and plating method using the same ', *U.S.Patent 5,674,374*, October 7., 1997

[72] Joseph, S; Girish, Phatak; Seth, T; Gurunathan, K; Amalnerkar DP; Kutty, TRN. "Lead free solder bumping by electroplating process for electronic packaging", *Proc. IEEE TENCON*, 2003, Vol. 4, 1367-71

[73] Shany Joseph; Girish, J; Phatak, K. Gurunathan, Tanay Seth, D.P. Amalnerkar And T.R.N. Kutty, 'Electrochemical co-deposition of ternary Sn-Bi-Cu films for solder bumping applications', *Journal of Applied Electrochemistry,* 2006, 36, 907–912,

[74] Allan da S. Taguchi, Fábio R. Bento and Lucia H. Mascaro, 'Nucleation and Growth of Tin-Zinc Electrodeposits on a Polycrystalline Platinum Electrode in Tartaric Acid', J. *Braz. Chem. Soc.,* 2008, 19, No. 4, 727-733,

[75] Guaus, E; Torrent-Burgue´s, J. 'Tin–zinc electrodeposition from sulphate–tartrate baths' *Journal of Electroanalytical Chemistry*, 2005, 575, 301–309

[76] Guaus*, E; Torrent-Burgue´s, J. 'Tin_/zinc electrodeposition from sulphate_/gluconate baths', *Journal of Electroanalytical Chemistry*, 2003, 549, 25-/36

[77] Chi-Chang Hu; Chun-Kou Wang; Gen-Lan Lee. 'Composition control of tin–zinc deposits using experimental strategies', *Electrochimica Acta*, 2006, 51, 3692–3698

[78] Bell; Jane; Heyer; Joachim; Hupe; Jürgen; Kalker; Ingo; Kleinfeld; Marlies. Process for the non-galvanic tin plating of copper or copper alloys', *US Patent 6821323* on November 23, 2004

INDEX

A

acetic acid, 37
acid, vii, 18, 20, 21, 24, 25, 26, 29, 37, 38, 39, 41, 44, 45, 48, 49, 50, 51
acidic, vii, 18, 20, 25, 28, 29, 31, 32, 37, 41, 44, 45, 49, 50, 52
additives, vii, 7, 8, 17, 19, 20, 24, 26, 31, 35, 37, 43, 44
adhesion, 4, 6, 53
amine, 24, 28, 45
amines, 26
anemia, 9
antimony, 14, 15
antioxidant, 19, 24, 30, 49
appetite, 9
aqueous solutions, 51
argon, 52
ascorbic acid, 26
atoms, 12
attachment, 2, 6

B

ban, 9
base, 44
baths, vii, 8, 17, 18, 19, 20, 21, 24, 25, 26, 28, 29, 30, 31, 32, 33, 35, 36, 37, 41, 44, 45, 46, 52, 53
biodegradability, 19
bismuth, 13, 48, 53

Blood lead level (BLL), 9
bonding, 2, 4, 28, 32, 33, 43
bonds, 9
by-products, 10

C

cadmium, 15
calcium, 41
carbon, 50
cathode polarization, 31
central nervous system, 9
chelates, 17, 29, 30, 31, 32, 33, 45
chemical, 6, 13, 25, 28, 49, 50
chemical properties, 6
chemical stability, 13
chemicals, 31
commercial, 18, 20, 21, 24, 26, 37, 45
compatibility, vii, 2, 28
compilation, 10
complexity, 3, 45
composition, vii, 1, 6, 7, 8, 12, 15, 17, 20, 21, 22, 24, 26, 27, 28, 29, 30, 32, 34, 35, 37, 38, 39, 41, 43, 45, 46, 51, 53
compounds, 37, 45
computer, 13
conductivity, 11, 16
constituents, 1, 7, 9, 17, 19, 21, 22, 26, 27, 30
contamination, 10

cooling, 9
copper, 6, 9, 12, 13, 20, 24, 44, 48, 49,
	50, 51, 54
corrosion, 11, 13, 15, 19, 20
corrosivity, 19
cost, 2, 3, 4, 6, 7, 9, 10, 11, 12, 13, 15,
	19, 47, 48
cracks, 12
crown, 50
crystals, 12
cyanide, 20, 25, 44, 49
cycles, 11, 15, 26
cycling, 11, 14, 15

D

dendrites, 28
deposition, vii, 2, 6, 7, 8, 15, 17, 18, 19,
	20, 21, 22, 24, 25, 26, 27, 28, 29, 30,
	31, 33, 37, 39, 40, 43, 44, 45, 46, 49,
	52, 53
deposition rate, 21, 24
deposits, 20, 26, 28, 30, 33, 38, 50, 53
derivatives, 45
diffusion, 4, 6
dizziness, 9
ductility, 14, 15, 20

E

effluent, 18, 20, 45
Egypt, 1
electrochemical deposition, 52
electrodeposition, 33, 34, 35, 50, 51, 52,
	53
electrodes, 50, 51
electrolyte, vii, 20, 21, 22, 24, 27, 28, 41,
	48, 49
electroplating, vii, 2, 6, 7, 8, 15, 17, 18,
	23, 24, 25, 28, 37, 43, 44, 47, 49, 51,
	52, 53
environment, 10, 11
equilibrium, 18
equipment, 7, 19
ethanol, viii, 30

evaporation, 2, 6, 7
exposure, 10, 14

F

films, vii, viii, 2, 20, 21, 25, 26, 27, 28,
	29, 30, 31, 33, 34, 35, 36, 37, 38, 39,
	40, 41, 43, 44, 46, 50, 51, 53
formaldehyde, 27
formation, 1, 9, 12, 13, 16, 18, 19, 21,
	24, 35, 46, 50, 51, 53
FTIR, 28, 52

G

gallium, 15
geometry, 7
glycol, viii, 24, 26, 27, 28
grants, 46
growth, 13, 21, 24, 28, 44
growth rate, 21, 24, 44
guidance, 46

H

hazards, 9, 43
health, vii, 9, 11, 43
height, 26
Heliotropin (HT), 31
high fat, 14
high strength, 14
history, vii, 7, 13
hydrofluoric acid, 18
hydrogen, 19, 33
hydroquinone, 19, 26, 49
hypertension, 9

I

ideal, 9, 10, 17, 44
images, 32
immersion, 4
independence, 45
India, 46
indium, 14, 15
industry, 1, 3, 5, 10, 11, 20, 48

infrastructure, 7
ingots, vii, 2
ingredients, 7
integrity, 9
interface, 4
ions, vii, 17, 18, 20, 28, 29, 32, 33, 35,
    37, 39, 40, 41, 45
iron, 1
issues, 2, 8, 13, 14, 18, 28

J

joints, 1, 3, 11, 13

K

ketones, 26
kinks, 12

L

landfills, 10
leaching, 10
lead, iv, vii, 1, 2, 4, 7, 9, 10, 11, 12, 13,
    14, 15, 18, 29, 37, 43, 44, 46, 48, 52
learning, 9
learning disabilities, 9
light, 36
linear Sweep voltametry (LSV), 31
loss of appetite, 9

M

manufacturing, iv, 7
mass, 1, 3, 28
materials, 2, 6, 10, 11, 47, 51
matrix, 4
matter, iv
mechanical properties, 2, 11, 12, 13, 14,
    16, 29, 41
melt, 6
melting, 2, 3, 7, 10, 11, 12, 13, 14, 15,
    20, 29, 41
melting temperature, 2, 14, 15, 29
mercury, 15
Mesopotamia, 1

metal ion, 17, 28, 29, 32, 33, 35, 39, 41,
    45
metal ions, 17, 28, 29, 32, 33, 35, 39, 41,
    45
metal salts, 17, 26, 45
metallurgy, 51
metals, vii, 12, 14, 18, 34, 39, 51
microstructure, 7, 17, 20, 21, 25, 27, 28,
    30, 31, 33, 35, 37, 44, 53
military, 10
modulus, 14, 15
moisture, 15
molecular weight, 27
molecules, 36
morphology, 19, 21, 25, 26, 27, 28, 38,
    41, 50

N

nervous system, 9
neutral, 41
nickel, 6, 47
nitrogen, 52
nonionic surfactants, 26

O

optimization, 46
oxidation, vii, 6, 13, 15, 18, 19, 20, 21,
    28, 33, 44, 45
oxidative damage, 50
oxygen, 18

P

Pacific, 52
palladium, 6
passivation, 19
permission, iv, 23, 24, 31, 32, 33, 34, 35,
    40
pH, 18, 21, 22, 25, 26, 29, 30, 37, 38, 41,
    44, 45
phenol, 18, 26
phenolphthalein, 19
photolithography, vii, 20
physical and mechanical properties, 11

pitch, 6, 7, 25
plasticity, 14
polarization, 30, 31, 32, 52
polymer, 44
porosity, 26, 30
potassium, 19, 29
precipitation, 28, 33, 39
prevention, 1, 10
promoter, 30, 32
pyrophosphate, 20, 25, 28, 29, 45, 52

R

reactions, 19, 25, 47
recommendations, iv
reliability, vii, 6, 7, 10, 11, 13, 14, 21, 34
requirements, 2, 7, 18
researchers, 10
resistance, 11, 14, 15, 20
restoration, 11
rheology, 7
rights, iv
room temperature, 14, 41
rotations, 26
routes, 10

S

salts, 7, 17, 21, 26, 30, 38, 39, 41, 45
SEM micrographs, 35
semiconductor, 48
sensitivity, 13
shear, 11, 16
shear strength, 11, 16
shock, 13
showing, 6
silver, 13, 22, 25, 28, 32, 37, 44, 50, 51, 52
smoothing, 21
smoothness, 30
society, 50
sodium, 19, 24, 41
solder balls, 4
solder bumps, 2, 4, 6, 21, 23, 24, 43, 47, 52

solder form has, vii
solubility, 14, 19
solution, 8, 18, 19, 20, 21, 29, 51
stability, 8, 13, 20, 21, 24, 25, 26, 29, 30, 33, 37, 38, 39, 45, 49
stabilizers, 19
stable complexes, 37
stress, 12, 18
structure, 4, 12, 15, 19
substrate, 4, 6
substrates, 4, 12, 25, 47
sulfate, 20, 45
sulfur, 28
sulfuric acid, 18, 20, 24, 25, 26, 29, 37, 44, 45, 50, 51
sulphur, 33
surface tension, 9, 20
surfactant, viii, 19, 22, 29, 33, 49
surfactants, vii, 19, 24, 26, 44
surplus, 33
susceptibility, 15
symptoms, 9

T

techniques, 3, 37
technologies, 5
technology, 3, 6, 48
temperature, 2, 7, 11, 13, 14, 15, 27, 28, 29, 41, 53
tensile strength, 11, 14, 15
tension, 9, 20, 22
texture, 28
tin, vii, 1, 9, 10, 12, 13, 18, 19, 26, 28, 48, 49, 50, 51, 52, 53, 54
Top Surface Metallurgy (TSM), 4
toxic effect, 9, 25
toxicity, 1, 15, 19, 20
transformation, 9
transmission, 11
transport, 50
treatment, 7, 18, 20, 45
triethanolamine (TEA), 29

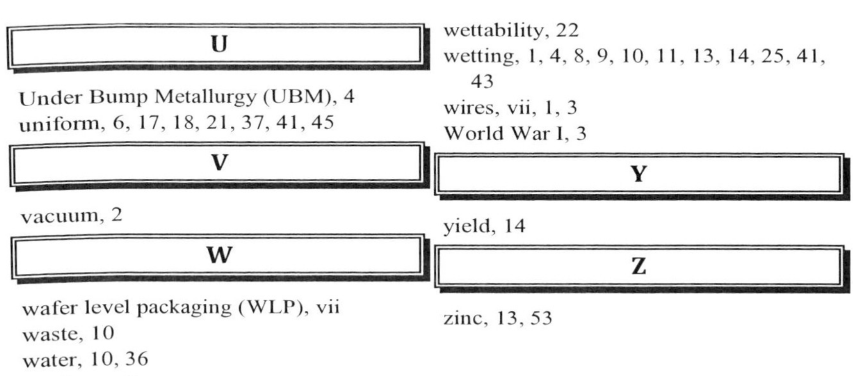

U

Under Bump Metallurgy (UBM), 4
uniform, 6, 17, 18, 21, 37, 41, 45

V

vacuum, 2

W

wafer level packaging (WLP), vii
waste, 10
water, 10, 36
weakness, 9

wettability, 22
wetting, 1, 4, 8, 9, 10, 11, 13, 14, 25, 41, 43
wires, vii, 1, 3
World War I, 3

Y

yield, 14

Z

zinc, 13, 53